D0546415

The Jessie and John Danz Lectures

THE JESSIE AND JOHN DANZ LECTURES

The Human Crisis, by Julian Huxley
Of Men and Galaxies, by Fred Hoyle
The Challenge of Science, by George Boas
Of Molecules and Men, by Francis Crick
Nothing But or Something More, by Jacquetta Hawkes
How Musical Is Man?, by John Blacking
Abortion in a Crowded World: The Problem of Abortion with Special Reference to India, by S. Chandrasekhar
World Culture and the Black Experience, by Ali A. Mazrui
Energy for Tomorrow, by Philip H. Abelson
Plato's Universe, by Gregory Vlastos
The Nature of Biography, by Robert Gittings
Darwinism and Human Affairs, by Richard D. Alexander
Arms Control and SALT II, by W. K. H. Panofsky
Promethean Ethics: Living with Death, Competition, and Triage, by Garrett Hardin
Social Environment and Health, by Stewart Wolf
The Possible and the Actual, by François Jacob
Facing the Threat of Nuclear Weapons, by Sidney D. Drell
Symmetries, Asymmetries, and the World of Particles, by T. D. Lee
The Symbolism of Habitat: An Interpretation of Landscape in the Arts, by Jay Appleton
The Essence of Chaos, by Edward N. Lorenz
Environmental Health Risks and Public Policy: Decision Making in Free Societies, by David V. Bates

The Possible and the Actual

FRANÇOIS JACOB

University of Washington Press
Seattle and London

Copyright © 1982 by François Jacob
First University of Washington Press paperback, 1994
Printed in the United States of America

All right reserved. No part of this publication may be reproduced or transmitted in any form or by any means, electronic or mechanical, including photocopy, recording, or any information storage or retrieval system, without permission in writing from the publisher.

Library of Congress Cataloging-in-Publication Data
Jacob, François, 1920–
The possible and the actual.
(The Jessie and John Danz lectures)
1. Evolution—Addresses, essays, lectures.
I. Title II. Series
QH371.J2 575 81-16452 AACR2
ISBN 0–295–97341–2 pbk.

The paper used in this publication meets the minimum requirements of the American National Standard for Information Sciences—Permanence of Paper for Printed Library Materials, ANSI Z39.48–1984. ∞

Cover illustration by Vesalius, from *De humani corporis fabrica* (1543)

Contents

The Jessie and John Danz Lectures

In October 1961, Mr. John Danz, a Seattle pioneer, and his wife, Jessie Danz, made a substantial gift to the University of Washington to establish a perpetual fund to provide income to be used to bring to the University of Washington each year "distinguished scholars of national and international reputation who have concerned themselves with the impact of science and philosophy on man's perception of a rational universe." The fund established by Mr. and Mrs. Danz is now known as the Jessie and John Danz Fund, and the scholars brought to the University under its provisions are known as Jessie and John Danz Lecturers or Professors.

Mr. Danz wisely left to the Board of Regents of the University of Washington the identification of the special fields in science, philosophy, and other disciplines in which lectureships may be established. His major concern and interest were that the fund would enable the University of Washington to bring to the campus some of the truly great scholars and thinkers of the world.

Mr. Danz authorized the Regents to expend a portion of the income from the fund to purchase special collections of books, documents, and other scholarly materials needed to reinforce the effectiveness of the extraordinary lectureships and professorships. The terms of the gift also provided for the publication and dissemination, when this seems appropriate, of the lectures given by the Jessie and John Danz Lecturers.

Through this book, therefore, another Jessie and John Danz Lecturer speaks to the people and scholars of the world, as he has spoken to his audiences at the University of Washington and in the Pacific Northwest community.

Preface

Some of the sixteenth-century books devoted to zoology are illustrated by superb drawings of the various animals that populate the earth. Certain ones contain detailed descriptions of such creatures as dogs with fish heads, men with chicken legs, or even women with three snake heads. The notion of monsters that blend the characteristics of different species is not surprising in itself: everyone has imagined or sketched such hybrids. What is disconcerting today is that in the sixteenth century these creatures belonged, not to the world of fantasies, but to the real world. Many people had seen them and described them in detail. The monsters walked alongside the familiar animals of everyday life. They were within the limits of the possible.

Let us not be ironic: we do exactly the same thing with our present-day science fiction books. The abominable animals that hunt the poor astronaut lost in a distant galaxy are products of recombination between organisms living on Earth. The creatures coming from outer space to explore our planet are depicted in the likeness of man. You can watch them emerging from their Unidentified Flying Objects: they are vertebrates, mammals without any doubt, walking erect. The only variations concern body size and the number of eyes. Generally, these creatures have larger skulls than humans, to suggest bigger brains, and sometimes one or two radioantennae on the head to suggest very sophisticated sense organs. The surprising point here is what is considered to be possible. It is the idea, one hundred and twenty years after Darwin, that if life occurs anywhere, it is bound to produce animals not too different from the terrestrial ones; and above all to evolve something like man.

The main interest of these monsters is that they show how a culture handles the possible and marks its limits. Whether in a social group or in an individual, human life always involves a continuous dialogue between the possible and the actual. A subtle mixture of belief, knowledge, and imagination builds before us an ever changing picture of the possible. It is on this image that we mold our desires and fears. It is to this possible that we adjust our behavior and actions. In a way, such human activities as politics, art, and science can be viewed as particular ways of conducting this dialogue between the possible and the actual, each one with its own rules.

This book deals with genes and men, with sex, aging, and molecules. Above all, it deals with the theory of evolution, its status as well as its content. For if the theory of evolution provides a framework without which there is little hope to understand where we come from and why we are as we are, it is important to define the limits beyond which it is used no longer as a scientific theory, but as a myth.

In recent years, I have given two lectures concerned with some aspects of these topics. One, delivered at the Weizmann Institute in Israel and at the University of California in Berkeley, was published in *Science* under the title "Evolution and Tinkering"; the other, delivered at the Académie de Chirurgie in Paris, was published in the *Journal de Chirurgie* under the title "Mon dissemblable mon frère." The invitation to deliver the Jessie and John Danz Lectures at the University of Washington gave me the occasion to collect and expand these discussions. To all those who had some responsibility for this invitation I extend thanks for having provided me the opportunity of writing this book and for their warm hospitality during my stay in Seattle.

The Possible and the Actual

"One can't believe impossible things" [Alice said].

"I dare say you haven't had much practice," said the Queen. . . . "Why, sometimes I've believed as many as six impossible things before breakfast."

—*Lewis Carroll*, Through the Looking Glass

MYTH AND SCIENCE

Theories pass. The Frog remains.

—*Jean Rostand,* Notebooks of a Biologist

Someday, perhaps, the physicists will be able to prove that nature could not work otherwise than it does. Someday, perhaps, they will build a theory showing that the actual world is the only possible one, that one cannot conceive of matter that is endowed wih properties different from what they are. Yet it seems difficult not to see some arbitrariness, some whimsicality, in the structure and functioning of our world. In a tale of my childhood, a fairy gives a young prince the following advice: "Blow the horn and the ogre's castle will fall down." In the Bible, Joshua runs seven times around Jericho, blowing his trumpet until the city walls fall. In both worlds, a causal relationship is clearly established between the blowing of an instrument and the falling of walls. This is the way it works. Things just happen to be that way. Relatively speaking, our physical universe also shows some arbitrariness. Nature also just happens to be as it is. It seems difficult, for me at least, to imagine a world in which one and one would not make two: this relation has a character of necessity, of inescapability, perhaps because it reflects the very way our brains work. On the other hand, it is possible to imagine a world in which physical laws would be different; in which ice, for instance, instead of rising would sink in water; or in which cutting the stem of an apple on a tree would cause the apple not to fall but to rise up through the air and out of sight.

Such speculation is even easier with the living world, which appears to be just one among an infinitude of possible situations. This is so, not only because the forms of living organisms could be very different, but also because of their functioning, of such features as death and reproduction. It is difficult to see any necessity in the fact that trees bear fruits, or that animals age, or in sex. How is it that, in the human body, reproduction is the only function to be performed by an organ of which an individual carries only one half so that he has to spend an enormous amount of time and energy to find another half?

In fact, sex is not a necessary condition for life. Many organisms have no sexuality and yet look happy enough. They reproduce by fission or budding and a single organism is sufficient to produce two identical ones. So how is it that *we* do not bud or divide? Why do most animals and plants have to be two in order to produce a third one? And why two sexes rather than three? What a source of new plots three sexes would provide for novelists, of new variations for psychologists, of new complications for lawyers. But perhaps this would be too much for us. Perhaps we could not cope with so many delights and problems. Let us content ourselves with our two sexes.

Every human culture accounts for the existence of the two sexes by myths that explain the origin of the world, of animals and of humans. In fact, there are only two ways of considering the genesis of the sexes, and mythologies have elaborated an infinite number of variations on these two themes. First, one can look at sexuality as a primary phenomenon, so to speak. The two sexes are then as old as the world itself. Before they existed, there could not be any life. The sexual duality reflects a cosmic duality. It expresses the two poles of the forces that are supposed to run the world, as observed in natural phenomena: day and night, sky and earth, water and fire. In the Tao, for instance, Yin and Yang, the male and female principles, are at the origin of every thing, every life, every motion. Similarly, in the Sumerian cosmogony, water, which represents the initial manifestation of life in the world, comes in two states: Apsu, fresh water, or the male principle; and Tiamat, salt water, or the female principle. The union of Apsu and Tiamat gives birth to Mummu, a kind of animate water which possesses mind and logos. Another variation is found in Egypt, where the primary divinity, Khoum, was one; but the first concern of the God was to produce a couple, Chou and Tefnout, who gave birth to mankind by the usual procedure of couples. Finally, in an interesting variation, the Veda presents the primeval couple as twins, Yami and Yama, and it is from a primitive act of incest that the human species is born.

But one may as well consider the sexual duality not as a primary but as a secondary phenomenon. What was originally created was one. Only afterward did it become two. The variations then concern the way the two sexes were formed, the event by which the original unity was broken. For the Upanishads, it is the God who, wanting to escape loneliness, resolves himself into two halves of opposite sex which then generate mankind. In other cultures, in contrast, sexual differentiation appears among beings that are neither quite gods, nor quite humans. In some tales of Zarathustra, for instance, Yima, the being created by God, is a kind of monster in which both sexes are united. This unity is only provisional, however, and Yima is soon sawed in two halves. A similar situation obtains in the famous account by Aristophanes in Plato's *Symposium*: at a time when sexuality was already working quite successfully among the gods and goddesses of Mount Olympus, what was to become mankind had not yet passed beyond the state of Androgynes. These spherical organisms were endowed with a bifacial head, four feet, four hands, four ears, and a double set of privates. They could move at full speed by rolling over and over. Ultimately, their strength and boldness began to worry Zeus, who decided to cut them into halves "as an egg with a horsehair," Plato specifies. Apollo was commissioned to perform surgery on the Androgynes and to sew them up so as to make mortals more modest and yet decent in appearance. Since that time, each of these halves has striven to join with another half which, for the Greeks, did not necessarily have to be of the opposite sex. Finally, another variation on the same theme is found in the Old Testament. The human being is created in his final, male aspect and not as a precursor monster. Subsequently Eve is produced from Adam. By dividing the unique, by cutting woman out of man, Genesis forces them to reconstruct the initial being in order to multiply.

Until the second part of the nineteenth century, science had very little to say about sex. Biologists could only try to describe it and to draw up an inventory of its multifarious aspects. Sex appeared as a fact to which, as Georges-Louis Leclerc de Buffon put it in the eighteenth century, "there is no other solution than the fact itself."[1] Only in the framework of the theory of evolution could sex secure scientific status. Only then could a new type of explanation be provided that was no longer based on origin, but on function. This function was suggested by Darwin himself as well as by August Weismann. It is, Weismann wrote, to produce "individual differences through which natural selection creates new species."[2] Although proposed at a time when the mechanism of heredity was still unknown, this theory has been amply supported by classical genetics and by molecular biology.

For modern biology, every living organism represents the expression of a program coded in its chromosomes. In organisms that reproduce asexually, for instance by fission, the genetic program is exactly copied out at every generation. Except for rare mutants, all individuals of a population are thus identical. Such populations can adapt only by the selection of these rare mutants under environmental conditions. In contrast, when sex has become a necessary condition for reproduction, every program is formed, not by the exact copy of a previous program, but by the reassortment of two different ones. As a result, every genetic program—every individual—becomes different from all others in the population with the exception of identical twins. Every child conceived by a given couple is the result of a genetic lottery. He is merely one out of a large crowd of possible children, any one of whom might have been conceived on the same occasion if another of the millions of sperm cells emitted by the father had happened to fertilize the egg cell of the mother—an egg cell which is itself one among many. And all these possible children would be as different from one another as the actual ones. If we go to all the trouble it takes to mix our genes with those of somebody else, it is in order to make sure that our child will be different from ourselves and from all our other children.

Sex is thus considered as a diversity-generating device. There remain many unanswered questions, concerning for instance the way sexuality arose in evolution, the relative advantages of various forms of parthenogenesis and hermaphroditism as compared with sexual reproduction, sex ratios, the importance, if any, of group selection, etc. Yet as emphasized by Ronald Fisher, Hermann Muller, and more recently George Williams and John Maynard Smith, the reshuffling of genetic material at every generation makes it possible to rapidly bring together favorable mutations which in asexually reproducing organisms would remain separate.[3] A population with sex can thus evolve faster than a population without it. In the long run, sexual populations will survive when asexual ones will die out. Furthermore, sexually reproducing individuals give rise to progeny with a wider range of phenotypes. In the short run, they have, therefore, a greater chance of producing some offspring of high fitness when the environment fluctuates. Thus sex provides a margin of safety against environmental uncertainty. It makes extinction less likely. It is an adaptation to the unforeseeable.

In some respects at least, myths and science fulfill a similar function: they both provide human beings with a representation of the world and of the forces that are supposed to govern it. They both fix the limits of what is considered as possible. In their modern form, the sciences were born at the end of the Renaissance, that is, at a time when Western man completely changed his relation to the world around him as he tried to reconstruct a universe that would be in ever better agreement with the evidence coming from sensory perception. From the Renaissance onward, Western art became radically different from all other art. With the invention of linear perspective and depth, of light and shade, the very function of painting was transformed in Europe within a few human generations: instead of symbolizing, it began to represent. A visit to a museum reveals a continuous series of efforts which, in many ways, recall those of science. From the Primitives to the Baroques, the painters never stopped improving their means for representing, for showing things and beings in the most faithful and convincing way. Playing with optical illusions allowed them to create an entirely new world, an open world in three dimensions. Between a Madonna by Cimabue, standing stark and stiff in her veils, rooted in a symbolic landscape, and Titian's mistress, lying naked on a bed, the same kind of break can be observed as that we see between the closed world of the Middle Ages and the infinite universe which began to reveal itself after Giordano Bruno. For painting and astronomy, each in its own domain, these changes reflected a general transformation associated with the political conquest of the globe, a transformation by which Western man completely renewed his mental models of the world. Another relation appeared between the imaginary and the real, between the possible and the actual. From the thirteenth century to the classical age in Europe, not only was pictorial representation substituted for symbolization, but also action for prayer, history for chronicle, drama for mystery, novel for tale, polyphony for monody, and scientific theory for myth. Yet it is probably the structure of the Judeo-Christian myth that made modern science possible. For Western science is founded on the monastic doctrine of an orderly universe created by a God who stands outside of nature and controls it through laws accessible to human reason.

It is probably a need of the human brain to have a representation of the world that is unified and coherent. A lack of unity and coherence frequently results in anxiety and schizophrenia. It seems only fair to say that, as far as coherence and unity are concerned, mythical explanation often does better than scientific explanation. For science does not aim at reaching a complete and definitive explanation of the whole universe. It proceeds by detailed experimentation on limited areas of nature. It looks for partial and provisional answers for certain phenomena that can be isolated and well defined. Other systems of explanation, whether magical, mythic, or religious, encompass everything. They apply to every domain. They answer every possible question. They account for the origin, the present state, and even the ultimate fate of the universe. One may disagree with the type of explanation offered by myth or magic. One cannot, however, deny them unity and coherence since, with no hesitation, they can answer every question and explain everything by one simple a priori argument.

At first glance, science appears to be less ambitious than myth in the type of question it poses and the kind of answer it seeks. In fact, the beginning of modern science can be dated from the time when such general questions as "How was the universe created? What is matter made of? What is the essence of life?" were replaced by more modest questions like "How does a stone fall? How does water flow in a tube? What is the course of blood in the body?" Curiously enough, this substitution had a quite unexpected result. While asking general questions led to very limited answers, asking limited questions turned out to provide more and more general answers. This is still valid for present-day science. The capacity to judge what problems are ripe for analysis, to decide when it is useful to reinvestigate old territory, to reexamine questions that once were considered as solved or insoluble, remains one of the most important qualities of a scientist. Creativity in science often corresponds to sure judgment in this domain. It is characteristic of the inexperienced young scientist and of the amateur that they do not content themselves with limited problems but want to tackle what they consider as general issues.

By the very way it proceeds, the scientific method has led to a fragmentation of the world view. Each branch of science has its language, its techniques, its domain. It is not necessarily connected with neighboring fields. Scientific knowledge often appears to consist of isolated islands. Important advances may sometimes come from new generalizations that unify what heretofore appeared as separate fields. Thus thermodynamics and mechanics were unified through statistical mechanics; as were optics and electromagnetism through Maxwell's theory of the electromagnetic field, or chemistry and atomic physics through quantum mechanics. Despite such generalizations, however, large gaps remain in scientific knowledge, some of which will probably not be bridged for a long time, if ever.

In their attempt to perform their function and to transform the chaos of the world into order, myths and scientific theories operate on the same principle. The object is always to explain visible events by invisible forces, to connect what is seen with what is assumed. A thunderstorm can be considered as the expression of Zeus's anger or as an electrostatic phenomenon. A disease can be viewed as the result of a spell cast on the patient or of a viral infection. In any case, however, a phenomenon is considered to be explained if it can be viewed as the visible effect of some hidden cause related to the whole network of invisible forces that are supposed to govern the world.

Whether mythic or scientific, the view of the world that man builds is always largely a product of his imagination. For, in contrast to what is frequently believed, the scientific process does not consist merely in observing, in collecting data and deducing a theory from them. One can watch an object for years without ever producing any observation of scientific interest. Before making a valuable observation, it is necessary to have some idea of what to observe, a preconception of what is possible. Scientific advances often come from uncovering some previously unseen aspect of things, not so much as a result of using some new instrument, but rather of looking at objects from a new angle. This look is necessarily guided by a certain idea of what so-called reality might be. As pointed out by Peter Medawar, it always involves a certain representation of the unknown, that is, of what is beyond that which one has logical or experimental reason to believe.[4] Scientific investigation begins by inventing a possible world, or a small piece of a possible world. So also begins mythical thought. But the latter stops there. Having constructed what it considers not only as the best world but as the only possible one, it easily fits reality into its scheme. Every fact, every event, is interpreted as a sign that is emitted by the invisible forces controlling the world, and therefore that proves their existence and importance. For scientific thought, in contrast, imagination is only one aspect of the game. At every step, it has to meet with criticism and experimentation in order to determine what might reflect reality and what is mere wishful thinking. For science, there are many possible worlds; but the interesting one is the world that exists and has already shown itself to be at work for a long time. Science attempts to confront the possible with the actual. It is the means devised to build a representation of the world that comes ever closer to what we call reality.

One of the main functions of myths has always been to account for the bewildering and meaningless situation in which man finds himself in the universe. They aim at providing a meaning to the disconcerting vision that man gains from experience alone, and at raising his confidence in life despite sufferings, mishap, and misery. The view of the world offered by myths is thus intermingled with everyday life and human feelings. Furthermore in a given culture, a myth which is repeated under a similar form, with similar words, from generation to generation, is not just a tale from which inferences can be drawn about the world. A myth has moral content. It carries its own meaning. It secretes its own values. In a myth, human beings find their law, in the highest sense of the word, without having to search for it. Even if they are looking for it, they can find a law neither in the conservation of mass and energy, nor in the primordial soup of evolution. In fact, science is an attempt to free investigation and knowledge from human emotional attitudes. The scientist seeks to remove himself from the world he is endeavoring to understand. He tries to step back, so to speak, to the position of a spectator who is not part of the world under study. By this trick, the scientist hopes to investigate what he considers to be "the real world around him." This so-called "objective world" thus becomes devoid of mind and soul, of joy and sadness, of desire and hope. In brief, this scientific, or "objective," world is completely dissociated from our familiar world of everyday experience. This attitude underlies the whole network of knowledge developed by Western sciences since the Renaissance. Only with the advent of microphysics has the border between observer and observed become somewhat blurred. The objective world is no longer that objective.

It has been an ever recurrent problem in the natural sciences to get rid of anthropomorphism, that is, to avoid endowing various entities with human properties. In particular, the purposive activity of man has long been considered to be the universal model for what appears as goal-directed processes in nature. Because of the obviously purposive aspects of living beings, their properties and their behavior, the living world has constituted the favorite target of final causes. In fact, the main argument for the existence of God has long been the famous "argument from design." It was developed in particular by William Paley in his *Natural Theology*, published a few years before the *Origin of Species*, and it goes as follows.[5] If you find a watch, you will scarcely doubt that it was designed by a watchmaker. Similarly, if you consider a complex organism with all its purposeful organs, you cannot escape the conclusion that it was designed by the will of a Creator. For it would simply be absurd, Paley says, to suppose that the eye of a mammal, for example, with the precision of its optics and its geometry, could have developed by mere chance.

There are two levels of explanation for goal-directed processes in living beings and they have frequently been confused. One deals with the individual organism, of which many morphological, biochemical, and behavioral properties clearly appear as goal-oriented. This applies, for instance, to the various phases of reproduction, to the development of the embryo, digestion, respiration, the search for food, escape from predation, migration, and so forth. This kind of preestablished design that is manifest in every living organism is not found among inanimate objects. It was, therefore, long considered to result from a special agent, some vital force escaping the laws of physics. Only in the last decades has a mechanistic interpretation of the activities observed in a living organism been considered to be compatible with its properties and behavior. In particular, the paradox came to an end when molecular biology borrowed from the theory of information the concept and term of "program" to describe the genetic information of an organism. Accordingly, the chromosomes of the fertilized egg are assumed to contain, coded in the DNA, the genetic blueprint that is considered to direct the development of the future organism, its activities and behavior.

The second level of explanation concerns not the individual organism but the whole living world and its present state. That is the area where the idea of special creation, the idea that each species was separately designed by a Creator, was demolished by Darwin. Against the argument from design, Darwin showed that a combination of certain mechanisms could actually mimic design; that it was possible to explain what appeared to be goal-directed activities by the chance variation of characteristics, followed by natural selection. When Darwin proposed this new way of looking at the living world, the mechanism underlying heredity was still completely unknown. Yet Darwin's model has been supported first by Mendelian genetics and later by molecular biology, which provided a genetic and biochemical basis for understanding reproduction and variation. Biologists have thus progressively elaborated a reasonable, albeit still incomplete, picture of what is considered to be the main force driving evolution of the living world, namely natural selection.

Natural selection can be viewed as the result of two constraints imposed on every living organism: first, the requirement for reproduction, which is fulfilled through genetic mechanisms carefully adjusted by such special devices as mutations, recombination, and sex to produce organisms similar, but not identical, to their parents; and second, the requirement for a permanent interaction with the environment, because a living being represents what thermodynamicists call an "open system" that can subsist only by virtue of a constant flux of matter, energy, and information. The first of these two factors generates random variations and produces populations in which all individuals are different. The interplay of the two factors results in differential reproduction and consequently in populations that evolve progressively as a function of environmental circumstances, behavior, and new ecological niches. Natural selection, however, does not act merely as a sieve eliminating detrimental mutations and favoring the reproduction of beneficial ones, as is often suggested. In the long run, it integrates mutations and orders them into adaptively coherent patterns adjusted over millions of years and over millions of generations as a response to environmental challenges. It is natural selection that gives direction to changes, orients chance, and slowly, progressively produces more complex structures, new organs, and new species. The Darwinian view has, therefore, an inescapable conclusion: the actual living world, as we see it today, is just one among many possible ones. Its present structure results from the history of the earth. It might well have been very different; and it might even not have existed at all!

The Paley versus Darwin argument, the opposition between special creation and natural selection, may be taken as an example of the debate between what Joshua Lederberg has called instructionism and selectionism.[6] While Darwin's model is selectionist, the theistic theory may be regarded as instructionist. For the Creator acts as a sculptor who, by design, molds matter and instructs it what shape to take; or as an information engineer who writes a program and instructs the computer what operation to perform. All mythologies use the human model of teaching and creating. All have an anthropomorphic attitude and are instructionist. The importance of Darwin's solution was to explain by the selection of already formed structures something that very much looks like instruction.

The selectionism versus instructionism debate has permeated the whole of biology. Its most famous aspect concerns the inheritance of acquired characteristics—the idea that living organisms receive from the environment, from the repetition of certain actions, genetic instructions that are passed on from one generation to the next. In this Lamarckian view of heredity, the genetic equipment of living organisms is supposed to learn from the environment just as the human brain does. Instructionism results from attributing to biological processes properties that belong to the mental processes of humans. This explains our overwhelming tendency to believe in an instructive or Lamarckian theory of heredity and evolution. Already the Bible was Lamarckian, as shown, for instance, by a beautiful experiment performed by Jacob. To avoid troubles with his father-in-law, Jacob decided to label his own sheep. So he "peeled white streaks in rods of green poplar and set them in the watering troughs."

When the animals came to drink, they "conceived before the rods and brought forth cattle ringstreaked, speckled and spotted." Down through the ages, a great many similar experiments were carried out, although not always so successfully.

Until the nineteenth century, the instructionist nature of heredity was not even questioned. The first anti-instructionist experiment was performed in the 1880s by August Weismann, who wanted to support his theory of the independence of the soma and the germen.[7] In order to prove that the germ cells are sheltered from the misfortunes of the body and that experience does not teach heredity, Weismann cut off the tails of newborn mice during many successive generations. After such treatment repeated many times from parents to offspring, he noted with pleasure that the young mice still grew normal tails. This experiment, however, did not appear very convincing. Only at the beginning of this century was Lamarckism refuted when it was seen to be totally incompatible with the properties of the genes and their mutations. Each carefully designed and strictly executed experiment planned to evaluate genetic instructionism has shown it to be wrong. For modern biology, there is no molecular mechanism enabling instructions from the environment to be imprinted into DNA directly, that is, without the roundabout route of natural selection. Not that such a mechanism is theoretically impossible. Simply it does not exist.

Yet although the inheritance of acquired characteristics has thus been removed from what biology considers the actual world, it has proved to be especially hard to kill in the mind, not only of laymen, but also of some biologists. Many experiments to save Lamarckism have continued to be performed, and some still are. Instructionist theories of heredity have remained a kind of *terra electa* for attempts to impose wishful thinking on the actual world, as illustrated by the Lysenko affair, and by a number of fakes, the most famous of which has been described at length by Arthur Koestler in his novel *The Case of the Midwife Toad*. The rule of the whole game in science is not to cheat—not to cheat with ideas, nor with facts. This is a rational as well as a moral commitment. The one who cheats in science is simply missing the point. He defeats himself. He commits suicide. Scientific fakes are indeed both surprising and interesting. They are surprising because on important issues it seems simply childish to think that the fraud would remain unnoticed for more than a short while; the cheater must therefore firmly believe that the claim, whose evidence he is merely faking, is in fact part of the actual. Fakes are also interesting because they range from deliberate forgery to slight, and sometimes even unconscious, deviations from the normal behavior of scientists. They touch, therefore, on psychological and ideological aspects of science and scientists. They may contribute to an understanding of the biases which, during a given period, impede scientific endeavor. In this sense, fakes belong to the history of the sciences.

Instructionist hypotheses were also considered at first to account for the specific properties of certain proteins. Many bacteria, for instance, are able to ferment a variety of sugars. However, they produce the enzymatic activities required to metabolize a particular sugar only when they are grown in a medium containing that sugar. It was long assumed that the sugar was bringing information to the bacterium; that it was in some way instructing the protein how to fold in order to have that particular enzymatic activity. Yet when bacteria became accessible to genetic analysis, this instructionist explanation proved to be wrong. The sugar molecule acts merely as a signal to initiate the synthesis of the protein. It selects from the genetic repertoire and activates the gene coding for that protein, the structure and activity of which remain completely independent of the sugar.

A similar situation obtained for the production of antibodies, the protein molecules that are produced by vertebrates in response to the injection of an antigen, i.e., a molecular structure that is not a constituent of the body and is considered by it as "foreign." A mammal can thus synthesize some 10^7 or 10^8 different antibodies, each one capable of "recognizing" a particular molecular structure, even if it has never seen this structure before. Because of this enormous number and of the impossibility of having in the germ line one particular gene for every possible antibody, the immune system has long remained one of the favorite areas for instructionist theories. The antigen was supposed to teach the antibody molecule what conformation to take in order to bind it. It is now clear that the system does not work in that way, but according to a more subtle mechanism. The production of antibodies is not an instructive but a selective mechanism. It involves not Lamarckism but Darwinism. However bizarre an antigen may be, an immunological response always represents a selection from a repertoire of preexisting structures, the activation of genetic information already present in the lymphoid cells, and not some kind of education of the cell by the molecular structure of the antigen.

There remains one domain in which the instruction versus selection argument has not been settled as yet: the nervous system. Very little is known about the way neurons become connected during embryogenesis, about the direct or indirect role played by the genes in the establishment of the "wiring diagram," or about the learning process. As in the immune system, the enormous number of synapses and the impossibility of having in the germ cell a particular gene for each synapse have led neurobiologists to assume that synapses become established through rather flexible, nongenetic mechanisms. The brain is by definition the domain of instructionism, and selectionist theories are badly received there because of the compelling argument that "*Macbeth* cannot be biologically prewired in the head of the child who learns it." The problem, however, is not a matter of words or ideas, but of synapses. Already several decades ago, it was suggested that a huge excess of synapses might become established during embryonic development, with learning then resulting from the selection of certain synapses and their combination in functional circuits, while unused synapses would disappear. It will probably take time before the instructionist versus selective nature of the learning process can be resolved.

Originally the theory of evolution was based on morphological, embryological, and paleontological data. During the present century, it has been strengthened by a series of results harvested by genetics, biochemistry, and molecular biology. All the information coming from various fields is combined in what is known as modern Darwinism. The traces of evolution can be found in each of our cells, in each of our molecules. It has today become virtually impossible to account for the tremendous amount of data accumulated during the last few decades without a theory very close to modern Darwinism.

The chance that this theory *as a whole* will someday be refuted is now close to zero.

Yet we are far from having the final version, especially with respect to the mechanisms underlying evolution. Genetics considers the organism on two quite distinct levels. One level deals with visible characteristics, morphology, functions, and behavior—what are called *phenotypes*. The other level deals with hidden structures, the state of the genes—so-called *genotypes*. These are in fact two quite different worlds. The former is concerned with the observation of actual organisms, the latter with explanation in terms of possible combinations of genes. Although genes control characteristics, the link between these two worlds has been made precise only for very simple traits. Only in such systems as blood groups or enzyme defects is it possible to establish a correlation between a given gene and its product, between genotype and phenotype. In most instances, however, the situation is much more complex. A single gene is often involved in the expression of many characteristics, and one characteristic may be controlled by many genes which we are not able to identify. Furthermore, we are still far from knowing all the genetic mechanisms underlying evolution, as is shown by recent findings about chromosome structure, breaks, and rearrangements. Virtually all biologists today believe in modern Darwinism. Some, however, think of evolution in terms of organisms, others in terms of molecules, and others in terms of statistical abstractions. There are still several possible ways of looking at evolution, its tempo and mechanism.

The device designed by Darwin to counteract Paley's argument from design was adaptation. This concept is at the center of the evolutionary representation of the world. It is inextricably linked with the present theory concerning the origin of the living world. Life is supposed to have originated in the so-called "primordial soup" resulting from chemical evolution, in the form of a molecular complex able to use some ingredients of the organic soup to reproduce itself with a possibility of variation. Natural selection could then operate. These early organisms increased their reproductive efficiency and began to diversify. One branch, which we call plants, succeeded in feeding directly on sunlight. Another branch, called animals, became able to use the biochemical capacity of the plants, either directly by eating them or indirectly by eating other animals that eat plants. Both branches found ever-new ways of living under ever-diversified environmental conditions. Subbranches appeared and sub-subbranches, each one becoming able to live in a particular environment, in the sea, on the land, in the air, in the polar regions, in hot springs, inside other organisms, etc. This progressive ramification over billions of years has generated the tremendous diversity and adaptation that baffle us in the living world of today.

The mechanism derived by Darwin from reading Malthus involves the advantage of those individuals which, because of their physiological or behavioral properties, make better use of the available resources for their reproduction. It links the genetic system responsible for reproduction and variation with the environment in such a way that the latter influences the former, a process which, in the long run, mimics Lamarckism. Adaptation results from competition among *individuals*, either within species or among species. It represents an automatic device that makes use of genetic opportunities and orients chance along paths compatible with life in a given environment. For many biologists, every organism, every cell, every molecule, has been refined to the last detail by an adaptation process that has been incessantly in action over millions of years and millions of generations.

This complete faith in the absolute power of natural selection has dominated evolutionary thought during the last fifty years. It has, however, been recently criticized by several population geneticists who refuse to believe that every organism is, down to the last detail of its cells and its molecules, molded for the best by adaptation. As pointed out by George C. Williams some fifteen years ago, adaptation is a special and onerous concept that should be used only when and where necessary.[8] Otherwise the indiscriminate use of this concept might lead to considering the living world to be nearly as perfect as the one formerly attributed to divine creation. The procedure of dissecting organisms into discrete characteristics, into structures each of which is assumed to best fulfill a function, results in what Stephen Gould and Richard Lewontin have called a Panglossian universe, a universe similar to that of Voltaire's hero.[9] When he learned about the great Lisbon earthquake in which some 50,000 people died, Dr. Pangloss explained to his pupil Candide: "All this is a manifestation of the rightness of things since if there is a volcano at Lisbon it could not be somewhere else. For it is impossible for things not to be where they are since everything is for the best."[10]

In fact, adaptation is not a necessary component of genetic evolution. A population is evolving whenever its gene pool changes in the course of generations whether it be gradually or suddenly. Such statistical alteration in the relative survival of genetic elements does not obligatorily imply adaptation. It can merely reflect some chance effect at any level of the reproductive process. Obviously random evolution cannot explain why land animals have legs, birds have wings, and fish have fins. Yet a number of mechanisms are now known to be operative in evolution besides natural selection, for instance genetic drift, random gene fixation, indirect selection due to genetic linkage, and differential growth of organs. Many of these factors help to randomize the effects of natural selection and to produce structures that may well be of no use. The problem is to determine the relative weight of all these processes in evolution.

Most important in restricting possible changes of structures and functions are the constraints imposed, for instance, by the general body plan underlying related species, by the mechanical properties of their building materials, and above all by the rules controlling embryonic development. For it is during embryonic development that the instructions contained in the genetic program of an organism are expressed, that the genotype is converted into phenotype. It is mainly the requirements of embryonic development that, among all possible changes in genotype, screen the actual phenotypes. When I was a young boy, I often wondered why human beings do not have two mouths rather than one: one with a sense of taste, strictly reserved for pleasant food, and another, without taste, for unpalatable things; or even why humans do not have, instead of hair, some kind of chlorophyl cap, so as not to have to spend so much time and effort in the search for food. The answer, however, seems fairly straightforward. Although such attributes would perhaps make life nicer or easier, our organization plan is the same as that of our vertebrate ancestors; and our vertebrate ancestors had only one mouth and no chlorophyl. Not just any organism is possible.

It should now be clear that there will no longer be a single recipe or a single theory to explain the whole universe in all its details. Yet so great appears to be the need of the human mind for unity and coherence of explanation that any theory of some importance is liable to be overused and to slip into myth. If it is to cover a large domain, a theory must be both powerful enough to explain a variety of events and flexible enough to apply to a variety of circumstances. Yet an excess of flexibility may well turn power into weakness. For a theory that explains too much ultimately explains very little. Its indiscriminate use invalidates its usefulness and it becomes empty discourse. Enthusiasts and popularizers, in particular, do not always recognize the subtle boundary that separates a heuristic theory from a sterile belief; a belief which, instead of defining the actual world, can describe all possible worlds.

The conceptual structures elaborated by Marx and Freud, for example, have been subject to such distortion by overuse. Freud was able to convince himself, and after him an appreciable fraction of the Western world, of the role played by unconscious forces in human affairs. After that Freud—and even more so his followers—tried desperately to rationalize the irrational, to enmesh it in an inescapable net of causes and effects. An amazing assortment of devices, including complexes, dream interpretation, transference, sublimation, and so forth, made it possible to explain any overt aspect of human behavior as a result of some covert psychological lesion. As for Marx, he propounded what he called "historical materialism" as one of the main forces orienting human societies and driving mankind on the road of "progress." Again, the followers of Marx felt it necessary to account for every aspect of the sound and fury of history by the same universal argument. Every detail of human history thus becomes a direct effect of some economic cause.

A theory as powerful as Darwin's could hardly escape misuse. Not only could adaptation serve to fit any detail of any structure found in any organism, but the very success of the theory of natural selection in accounting for the evolution of the living world made it tempting to generalize the argument and shape it to explain any change at all occurring in the world. Similar systems of selection have thus been invoked to describe any kind of evolution, whether cosmological, chemical, cultural, ideological, or social. Such attempts, however, appear to be doomed to failure from the start, for natural selection represents the outcome of specific constraints imposed on every single living organism. It corresponds, therefore, to a mechanism fitting that particular level of complexity. The rules of the game differ at each level. New principles have to be worked out at each level.

Among scientific theories, the theory of evolution has a special status, not only because some of its aspects are difficult to test directly and remain open to several interpretations, but also because it provides an account of the history and present state of the living world. In this sense, the theory of evolution is often given a status similar to that of a myth—of a story giving the origins and therefore explaining the meaning and purpose of the living world as well as man's place in it. As discussed before, every culture, every society, appears to have a need for myths, including cosmological myths. It may well be that these myths contribute to the cohesion of a human group by giving its members the link of a shared belief in a common origin and ascent. This community of belief probably allows the group to distinguish itself from "others" and to define its own identity. Although stories told about human evolution have been arranged so as to oppose so-called "civilized" and "primitive" human populations, the unity of mankind as a species prevents the theory of evolution from playing such a role, except perhaps to permit humans to differentiate themselves from Martians. Furthermore while a myth contains some kind of universal explanation that gives human life its meaning and ethics its values, it is not clear that the theory of evolution can also fulfill this function, in spite of various attempts.

In a universe created by God, the world and its inhabitants were necessarily as they ought to be. Nature was in a way modeled on morals. With the advent of the theory of evolution, it became tempting to reverse this situation and to infer morals from knowledge about nature. From its very beginning, Darwinism found itself entangled in ideology. Evolution by natural selection was immediately used in support of various doctrines and, since there are no moral values in natural processes, it could just as well be painted in pink or in black and be claimed to uphold any thesis. For Marx and Engels, the evolution of species was pointed in the same progressive direction as, and in accordance with, the history of societies. For capitalist and colonialist ideologies, Darwinism was invoked as a scientific alibi to justify social inequalities and various forms of racism. Since the middle of the nineteenth century, repeated attempts have been made—and sociobiology represents the most recent one—to ground morals on ethologico-evolutionary considerations. The ability to adopt a moral code may indeed be viewed as an aspect of human behavior. It must, therefore, have been shaped by selective, evolutionary forces in the same way as, for instance, the ability to speak, what Noam Chomsky calls a "deep structure."[11] In this perspective, it is the task of biologists to explain how human beings have evolved their *capacity* to hold ethical beliefs. This, however, does not apply to the *content* of these beliefs. It is not because something is "natural" that it is "right." Even if there exist temperamental and cognitive differences between the two sexes—and this point needs to be specified—it would not make it "good" nor "right" to force men and women into different social roles. There is no more reason to seek an evolutionary explanation for moral codes than there would be to seek such an explanation for poetry or physics. And nobody has ever suggested a biological theory of mathematics.

In fact, the search for biological answers to questions of ethics represents a confusion between what Kant considered to be two quite distinct categories. It is driven by the ideology of scientism, the belief that the methods and insights of the natural sciences will account for all aspects of human activity. Such a belief underlies the equivocal terminology used by many sociobiologists, as well as some of their unwarranted suppositions and extrapolations from animal to human behavior. On the other hand, confusion between science and ethics also appears in the opposite attitude which leads certain scientists to reject some well-grounded points of sociobiology merely because these arguments could be used to justify social policies they dislike—as if the theory of evolution were something more than a hypothesis to be tested and progressively adjusted; as if it had come to embody a variety of prejudices, hopes, and fears about our society.

All these controversies raise some very serious questions, such as: Is it possible for biologists to elaborate a theory of evolution that is really free of ideological bias? Is it possible for an account of origins to function both as a scientific theory and as a myth? Is it possible for a society to define a set of values directly, without referring to such external powers as God or history which man himself has created and set over his own existence?

EVOLUTIONARY TINKERING

Blood . . . is the best possible thing to have coursing through one's veins.

—*Woody Allen,* Getting Even

In 1543, the very year Copernicus' work was published, another book appeared: Vesalius' *De humani corporis fabrica*. This was a book of a completely new kind, not so much by its subject, human anatomy, as by its style. For the first time, the human body was not simply described in terms that had been passed down from generation to generation. It was represented in a series of plates that combined the art of the painter with the knowledge of the physician to display the structures progressively revealed by the scalpel. The book was not merely concerned with certain anatomical regions, as were the works of Dürer, Michelangelo, or Leonardo. It dealt with the whole architecture of the human body in attitudes connected with everyday life. Nothing earlier had reached the nobility and precision of these plates. In one, for instance, a skeleton is seen standing in profile, slightly stooped, with one elbow casually resting on a large table. The background represents a miniature landscape, that combination of ruins, palaces, and dwarf trees that the Renaissance used to delineate perspective. The clarity and precision of the bones are due to a soft light that comes from the top right and accentuates the shades on the back of the skull and vertebrae. The attitude of the skeleton is rather relaxed, so as to give an impression of both nonchalance and meditation. The nonchalance comes from its lopsidedness: the skeleton has its full weight on its outstretched right leg, while the left knee is slightly bent, the foot resting only on the toes. As for the sense of meditation, it is given by the left arm, bent with the elbow resting on the table, and by the head leaning on the back of the hand in the pose of the Thinker. The most surprising feature, however, that which gives the plate such intensity, is the fact that the face is turned toward a second skull which the right hand is holding on the table. With its empty orbits, the skeleton is staring at the other face, symbolizing man's desire to study himself.

Renaissance art had indeed never been sparing of skeletons. Yet, although Vesalius' figures grinned as stiffly as Holbein's or Dürer's, they did not fulfill the same function. In bas-reliefs or paintings, the skeletons symbolized the fragility of human life. They reminded the viewer that, in the face of death, everyone is equal. They prefigured the Last Judgment. In Vesalius' plates, however, their role is quite different. The skeletons represented from all angles demonstrate in its entirety the architecture that supports the human body. They display the structures that carry the attachment sites of the muscles and therefore allow humans to move and to work. In spite of their blank eyes, Vesalius' skeletons express, not the fear of death, but the activity of life.

Again, in Vesalius' muscle Tabulae it is the whole human body, front and back, that is depicted against the background of a landscape. Again, these figures are shown in familiar attitudes, with a mixture of energy and dignity on their tormented faces. They first appear stripped of their skin, thus exposing the whole network of superficial vessels. In the following plates, the layers of muscles are removed one after the other. Every muscle is cut at the top and pulled down, so as to reveal whatever was hidden beneath it. The body then continues to decrease in thickness. Every incision discloses some new shape; every gap some new symmetry of lines. Progressively, the whole space of the body comes into view. As the layers of muscles are removed, however, the body loses in drive and dignity. It slowly breaks down, collapsing a little more at every page. Step by step, it becomes a kind of puppet, leaning against a wall. Finally it is merely an empty carcass, held up only by a gallows rope. The story told by Vesalius' Tabulae is a familiar one today. At the time, however, it was quite new. It reminds us that if Western man has succeeded in making himself an object of science, it is through his own corpse. To know one's own body, one first has to destroy it.

For the sixteenth century, the human body was a unique structure. It bore no resemblance to that of any other living creature. To dissect corpses, to investigate their smallest details was not just to emphasize man's uniqueness and define his differences from the animals. It was also to give thanks to God. For the human body, said Jean Fernel, is "the supreme form, the most perfect of all sublunar forms."[1] Thus, explained Ambroise Paré, anatomy leads "directly to knowledge of the Creator just as the effect leads to knowledge of its cause."[2] The objects investigated by anatomy, the structures disclosed by dissection, were thus studied for themselves. Their main interest lay in their forms, which together give the human body coherence, life, and beauty. Anatomy was, therefore, an activity for painters and sculptors as much as for physicians. For at that time, a disease was not believed to be as directly related to a body function as is assumed today. It had no anatomical support, no organic cause. Being a disorder of the body, a disease could only express some unbalance in the forces that give life to this body; an unbalance in the humors, or in the relation between the soul and the body, or even in the network of secret influences that, from all over the universe, focus on man and work on him. A stomachache did not result from a lesion in the abdomen, but from an excess of humor, or from the influence of a star, or was due to an atonement, a vengeance, a divine punishment.

At the end of the Renaissance, anatomy was thus a closed science. Only later, in the seventeenth and eighteenth centuries, would the knowledge of the living organism and its constituents become based on their relations: the relation between structure and function as expressed in the circulation of the blood and Harvey's physiology; the relation between structure and disease as explained by Morgagni's pathology; the relation between structures belonging to different organisms as seen in comparative anatomy. Only through such comparison of forms and structures, through the idea that their distribution in space actually reflects a distribution in time, would a theory of evolution become possible.

These early days of anatomy are interesting, not only because they belong to a fascinating period, but also because, in some respects, modern biology finds itself in a somewhat similar situation. A few decades ago, the properties of living organisms were shown to result from the characteristics and interactions of the molecular structures of which they are composed. Since then, biologists have been hunting for molecules and it is not an exaggeration to say that almost every day new molecules are extracted from one organism or another. When observing an unknown phenomenon, a gifted young man will attempt to detect the proteins that are involved, to purify them and determine their amino-acid sequences. If he is really gifted, he will fish out their structural genes and define their nucleotide sequences. Yet, however gifted he may be, it will take the young man—and the old one as well—several decades or even centuries before he is able to figure out how such a molecule happened to be in this organism to perform what he considers to be its particular function.

All this sounds very much like molecular anatomy. To rationalize the structures revealed by the scalpel, sixteenth-century anatomists had to invoke God's will. To rationalize the structures revealed by chromatography, twentieth-century molecular biologists invoke natural selection, that is, a strange mixture of chance and reproductive competition. As a result of this attitude, history is promoted to the rank of a major causal agent.

In our universe, matter is arranged in a hierarchy of structures by successive integrations. Whether inanimate or living, the objects found on earth are always organizations or systems. Each system at a given level uses as its ingredients some systems from the simpler level. Molecules are made of atoms, but the molecules found in nature or produced in the laboratory represent only a small fraction of all the possible interactions between atoms; at the same time, molecules may exhibit new properties such as isomerization or racemization that do not exist in atoms. At the upper level, cells are made of molecules, but again the set of molecules found in living organisms represents a very restricted range of chemical objects; furthermore, cells are capable of division while molecules are not. At the next level, the number of animal species amounts to several million. Yet, this is small relative to the number that could exist. All vertebrates are composed of a very limited number of cellular types, at most a few hundred, such as nerve cells, gland cells, and muscle cells. The great diversity of vertebrates results from differences in the arrangement, in the number and distribution, of these few hundred types. The hierarchy in the complexity of objects has therefore two characteristics: first, the actual objects that exist at each level represent a limitation or restriction of the total number of possibilities offered by the combinatorial capacity of the next simpler level; second, at each level, new properties may appear that impose new constraints on the system. These, however, are merely additional constraints. Those operating at any given level remain valid at all more complex levels. But in most instances, the statements of greatest importance at one level have no interest at the more complex ones. The ideal gas law is no less true for the objects of biology than for those of physics. It is simply irrelevant to the problems that interest biologists.

Complex objects, whether living or not, are produced by evolutionary processes in which two kinds of factors are involved: the constraints that, at every level, specify the rules of the game and define what is possible with those systems; and the historical circumstances that determine the actual course of events and control the actual interactions between the systems. The combination of constraints and history exists at every level, although in different proportions. Simpler objects are more dependent on constraints than on history. As complexity increases, history plays a greater part. But history always has to be introduced into the picture, even in physics. For the universe itself and the elements that compose it have a history. According to present theories, heavier nuclei are composed of lighter ones and ultimately of hydrogen nuclei and neutrons. The transformation of heavy hydrogen into helium occurs during the fusion process, which is the main source of energy in the sun as well as in the hydrogen bomb. Helium and all the heavier elements are thus the result of cosmological evolution. According to present views, the heavier elements are considered to be products of supernova explosions and they seem to be very rare. The earth and other planets of the solar system have thus been formed of rare material under conditions that appear to obtain rarely in the cosmos.

Obviously history takes on much greater importance for biological objects. Since only constraints, but not history, can be formalized, biology has a scientific status different from that of physics. Explanation in biology has a dual character. In the study of any biological system, at any level of complexity, two kinds of questions can be asked: "How does it work?" and "How did it come about?" The first question, which deals with the here and now interactions, has been the main concern of experimental biology for the last hundred years. This biology is strongly mechanism-oriented and has provided a number of answers in physiological, biochemical, or molecular terms. The second question, which deals with evolution, is probably deeper because it contains the first one. In most instances, however, the answers can hardly be other than more or less reasonable guesses. The modern theory of evolution has based the rules of the historical game on two constraints imposed on living organisms: reproduction and thermodynamics. Yet for the understanding of certain structural and functional aspects of living organisms, not only the rules but in some cases the actual details of the historical process may be of importance. For every single organism living today represents the last link of a chain uninterrupted over some three thousand million years. Living beings are indeed historical structures; they are literally creations of history.

In the same way that comparative anatomy attempted to define morphological and functional relationships between species, so comparative molecular anatomy attempts to sketch some of the paths followed by evolution, especially those which could not be marked in fossils. This can be exemplified by the analysis of such a protein as cytochrome c, which has yielded information on one of the most fascinating aspects of the development of life on earth, namely the way organisms have obtained, stored, and used energy.[3] Cytochrome c serves as an electron shuttle in the electron transport chain of photosynthesis or respiration. The amino-acid sequence and even, in several instances, the three dimensional structure of cytochrome c have been determined for a variety of species. These include microorganisms of widely different types, such as aerobic bacteria that can use either oxygen or nitrate for oxidation, as well as green or purple photosynthetic bacteria and blue-green algae. They also include higher organisms, whether animals harboring mitochondria or plants harboring both mitochondria and chloroplasts. In a large number of these organisms, the similarities among the cytochromes c are striking. Regardless of their origin or their metabolic functions, all these cytochromes appear to belong to one evolutionarily related family of proteins and to have a common molecular ancestor.

Such an analysis yields two kinds of information. On the one hand, the combination of data about cytochrome c with those concerning other molecules makes it possible to outline a phylogenetic tree summarizing the relations between photosynthetic and respiratory bacteria. One can thus figure out how some major steps in the evolution of energy metabolism occurred, such as: the passage of sulfur photosynthetic bacteria to blue-green photosynthetic bacteria carrying the familiar cycle for reducing carbon dioxide; the progressive replacement of strong reductants, such as hydrogen sulfide, by water; the formation of an oxidizing atmosphere; and the appearance of respiration.

On the other hand, the evolution of cytochrome c shows the interplay of constraints and history at the molecular level. In a molecule such as cytochrome c, physical and chemical constraints appear to be especially strong because of the requirements of the heme and of the electrons which have to migrate freely through one edge of the molecule. When, at some early stage in the development of life, the basic structure turned out to be efficient in electron transport, it was thereafter maintained with very little change from photosynthetic procaryotes up to cells of various eucaryotes, protists, fungi, plants, and animals. With many other proteins, less stringent requirements allow enough historical variations to make the structures widely different in various species. With cytochrome c, however, little room is left for historical diversification. A few changes in amino-acids are permitted only at certain places. Although the different molecules are all folded in an identical way and present a similar tertiary structure, they vary in length from 82 to 134 amino-acids. The main differences consist in the addition or deletion of loops of chain at the surface of the molecule. This does not tell us much about the historical events that affected the molecule in the course of evolution. It does say something, however, about the way evolution proceeds to create molecular types.

The action of natural selection has often been compared to that of an engineer. This comparison, however, does not seem suitable. First, in contrast to what occurs during evolution, the engineer works according to a preconceived plan. Second, an engineer who prepares a new structure does not necessarily work from older ones. The electric bulb does not derive from the candle nor does the jet engine descend from the internal combustion engine. To produce something new, the engineer has at his disposal original blueprints drawn for that particular occasion, materials and machines specially prepared for that task. Finally, the objects thus produced *de novo* by the engineer, at least by the good engineer, reach the level of perfection made possible by the technology of the time. In contrast, evolution is far from perfection, as was repeatedly stressed by Darwin, who had to fight against the argument from perfect creation. In the *Origin of Species,* Darwin emphasizes over and over again the structural and functional imperfections of the living world. He always points out the oddities, the strange solutions that a reasonable God would never have used. One of the best arguments against perfection came from extinct species. While the number of species presently living in the animal kingdom can be estimated to be a few million, the number of extinct ones has been evaluated by George Gaylord Simpson at about five hundred million.[4] This means that some 99 percent of all species that once lived on earth have disappeared at some time or another.

In contrast to the engineer, evolution does not produce innovations from scratch. It works on what already exists, either transforming a system to give it a new function or combining several systems to produce a more complex one. Natural selection has no analogy with any aspect of human behavior. If one wanted to use a comparison, however, one would have to say that this process resembles not engineering but tinkering, *bricolage* we say in French. While the engineer's work relies on his having the raw materials and the tools that exactly fit his project, the tinkerer manages with odds and ends. Often without even knowing what he is going to produce, he uses whatever he finds around him, old cardboards, pieces of string, fragments of wood or metal, to make some kind of workable object. As pointed out by Claude Levi-Strauss, none of the materials at the tinkerer's disposal has a precise and definite function.[5] Each can be used in different ways. What the tinkerer ultimately produces is often related to no special project. It merely results from a series of contingent events, from all the opportunities he has had to enrich his stock with leftovers. In contrast with the engineer's tools, those of the tinkerer cannot be defined by a project. What can be said about any of these objects is just that "it could be of some use." For what? That depends on the circumstances.

In some respects, the evolutionary derivation of living organisms resembles this mode of operation. In many instances, and without any well-defined long-term project, the tinkerer picks up an object which happens to be in his stock and gives it an unexpected function. Out of an old car wheel, he will make a fan; from a broken table a parasol. This process is not very different from what evolution performs when it turns a leg into a wing, or a part of a jaw into a piece of ear. This point was already noticed by Darwin and discussed in the book he devoted to the fertilization of orchids, as pointed out by Michael Ghiselin.[6] Darwin showed how new structures are elaborated out of preexisting components, which initially were in charge of achieving a given task but became progressively adapted to different functions. For instance, the glue that originally held pollen to the stigma was slightly modified to affix pollen masses to the body of insects, thus allowing cross-fertilization by insects. Likewise, many structures that make no sense as features subservient to some end and which, according to Darwin, look like "bits of useless anatomy," are readily explained as remnants of some earlier functions. So, concludes Darwin, "if a man were to make a machine for some special purpose but were to use old wheels, springs and pulleys, only slightly altered, the whole machine, with all its parts, might be said to be specially contrived for that purpose. Thus throughout nature almost every part of each living being has probably served, in a slightly modified condition, for diverse purposes, and has acted in the living machinery of many ancient and distinct specific forms."

Evolution proceeds like a tinkerer who, during millions of years, has slowly modified his products, retouching, cutting, lengthening, using all opportunities to transform and create. The formation of a lung in terrestrial vertebrates, as described by Ernst Mayr, provides a clear example of this process.[7] Lung development started in certain freshwater fishes living in stagnant pools lacking oxygen. They adopted the habit of swallowing air and absorbing oxygen through the walls of the esophagus. Under such conditions, enlargement of the surface area of the esophagus conferred a selective advantage. Diverticula of the esophagus appeared and, under continuous selective pressure, enlarged into lungs. Further evolution of the lung was merely an elaboration of the same theme: enlarging the surface for oxygen uptake and vascularization. Making a lung with a piece of esophagus sounds very much like making a skirt with a piece of Granny's curtain.

When different engineers tackle the same problem, they are likely to end up with very nearly the same solution: all cars look alike, as do all cameras and all fountain pens. In contrast, different tinkerers interested in the same problem will reach different solutions, depending on the opportunities available to each of them. This variety of solutions also applies to the products of evolution, as is shown, for instance, by the diversity of eyes found throughout the living world. The possession of light receptors confers a great advantage under a variety of conditions. During evolution, many types of eyes appeared, based on at least three different principles: the lens, the pinhole, and multiple holes. The most sophisticated ones, like ours, are lens-based eyes, which provide information not only on the intensity of incoming light but also on the objects the light comes from, on their shape, color, position, motion, speed, distance, and so forth. Such sophisticated structures are necessarily complex. They can develop only in organisms that are already complex. One might suppose, therefore, that there is just one way of producing such a structure. But this is not the case. Eyes with lenses have appeared in molluscs and in vertebrates. Nothing looks so much like our eye as the octopus eye. Yet it did not evolve the same way. In vertebrates, the photoreceptor cells of the retina point away from light while in molluscs they point toward light. Among the many solutions found to the problem of photoreceptors, these two are similar but not identical. In each case, natural selection did what it could with the materials at its disposal.

Finally, in contrast with the engineer, the tinkerer who wants to refine his work will often add new structures to the old ones rather than replace them. This procedure is also frequently observed with evolution, as exemplified by the development of the brain in mammals. This development was not as integrated a process as, for instance, the transformation of a leg into a wing. It involved the addition, to the old rhinencephalon of lower vertebrates, of a neocortex which rapidly, perhaps too rapidly, took a most important part in the evolutionary sequence leading to man. Some neurobiologists, especially Paul McLean, consider that those two types of structures correspond to two types of functions and have not been completely coordinated or hierarchized.[8] The recent one, the neocortex, controls intellectual, cognitive activities while the old one, derived from the rhinencephalon, controls emotional and visceral activities. The old structure, which in lower mammals was in total command, has in men been relegated to the department of emotions, and constitutes what McLean calls the "visceral brain." Perhaps because development is so prolonged and maturity so delayed in man, this center maintains strong connections with lower autonomic centers and continues to coordinate such fundamental drives as obtaining food, hunting for a sexual partner, or reacting to an enemy. This evolutionary procedure—the formation of a dominant neocortex coupled with the persistence of a nervous and hormonal system partially, but not totally, under the rule of the neocortex—strongly recalls the way the tinkerer works. It is somewhat like adding a jet engine to an old horse cart. No wonder accidents occur.

It is probably at the molecular level that the tinkering aspect of evolution is the most apparent. What characterizes the living world is the basic unity that underlies its tremendous diversity. The living world contains bacteria and whales, viruses and elephants, organisms living in polar areas at −20°C and others living in hot springs at 70°C. All these creatures, however, exhibit a remarkable unity of structure and function. Similar polymers fulfill similar functions. The genetic code is the same and the translating machinery very nearly so. The same coenzymes mediate similar reactions. Many metabolic steps remain essentially the same from bacteria to man. Undoubtedly, for life to emerge, a number of new molecular structures had first to be produced. All those molecules out of which every living organism is built had to appear during chemical evolution in prebiotic times and at the beginning of biological evolution. But once life had started in the form of some primitive, self-producing organism, further evolution had to proceed mainly through alteration of already existing compounds. As new proteins appeared, new functions could develop. But these new proteins could only be mere variations on previous themes. A sequence of a thousand nucleotides codes fo a medium-sized protein. The probability that a functional protein could appear *de novo*, merely by random association of amino-acids, is practically zero. In organisms as complex and integrated as those that were already living a long time ago, the creation of entirely new nucleotide sequences could not be of any importance in the production of new information. The appearance of new molecular structures during much of biological evolution must, therefore, have resulted from the alteration of existing ones. This can be realized, for instance, by gene duplication. When a gene exists in more than one copy in a cell or a gamete, it is released from the constraints imposed by natural selection. Mutations can then accumulate more or less freely and result in a new structure. This procedure has been widely used in evolution as is shown by the existence of families of very similar proteins controlled by sets of genes that derive from a common ancestor, such as the family of the globins or that of the major histocompatibility antigens.

When DNA sequences are compared, large segments of genetic information turn out to be homologous, not only in the same organism, but also among different organisms, even among those that are phylogenetically distant. Similarly, as more is known about amino-acid sequences in proteins, it appears not only that proteins fulfilling similar functions in different organisms frequently have similar sequences, but also that proteins with different functions in the same organism often exhibit segments in common. As if structural genes coding for amino-acid sequences in proteins had been formed during evolution by the combination and permutation of smaller DNA segments.

Such a reassortment of DNA stretches, which must be invoked as the origin of new proteins during evolution, is exemplified by what is observed today in one particular aspect of embryonic development in mammals: the production of antibodies. As already mentioned in the previous chapter, a mammal can produce several millions or tens of millions of different antibodies, a number far greater than the number of structural genes in the mammalian genome. Actually a small number of genetic segments is used, but the diversity is generated during the development of the embryo by the cumulative effect of different mechanisms operating at three levels. First, at the *cell* level, every antibody-forming cell produces only one type of antibody, the total repertoire of antibodies in the organism being formed by the whole population of such cells. Second, at the *protein* level, every antibody is formed by the association of two types of protein chains, heavy and light; each of these chains can be sampled from a pool of several thousand and their combinatorial association generates a diversity of several million types. Third, at the *gene* level, every gene coding for an antibody chain, heavy or light, is prepared during embryonic development by joining several DNA segments, each one sampled from a pool of similar but not identical sequences. This combinatorial system allows a limited amount of genetic information in the germ line to produce an enormous number of protein structures with different binding capacities in the soma. This process clearly illustrates the way nature operates to create diversity: by endlessly combining bits and pieces.

Although the phylogenetic creation of new genes cannot offer the same degree of precision and efficiency as this ontogenic mechanism of antibody formation, similar principles might well have been involved. It seems likely that new genes have arisen through some process of joining preexisting DNA fragments at random. As a matter of fact, it seems necessary to assume that such a mechanism for joining stretches of DNA goes far back in evolution, for early organisms could not start off by producing large proteins. In all likelihood, it all began with small stretches of thirty to fifty nucleotides produced by chemical evolution, each able to code for some ten to fifteen aminoacids. Only secondarily could such stretches be joined at random by some ligating process to form longer protein chains, some of which turned out to be useful, and were thenceforth selected. If this view is correct, one should expect to find more and more fragments of DNA sequences that are common to what look like unrelated genes. As the analysis of DNA and protein sequences proceeds, not only should new families and domains appear, but also subfamilies and sub-subfamilies. Once again, it is difficult to see how molecular evolution could have proceeded if not by turning old into new by knotting pieces of DNA together—that is, by tinkering.

Chromosomes have long been considered to be some kind of intangible and rather perfect structure containing exactly the amount of genetic information required for the production, form, and functioning of the organism. In the last few decades, however, this picture has been greatly modified by several findings. In addition to specific sequences coding for proteins, the DNA of higher organisms contains a large part, amounting sometimes up to more than 40 percent of the genome, of nonspecific DNA made of small, highly reiterated sequences. Even the coding sequences are frequently interrupted by a variable number of intervening sequences that are transcribed into RNA, but spliced out from RNA before translation. Furthermore the genome contains a class of genetic units known as "transposable elements," which can be inserted in, or excised from, a large number of sites in the host DNA where they can produce mutations, inversions, transpositions, and so forth. No function has yet been found for most of these noncoding sequences and their status is still under debate. For some biologists, the difficulty of accepting structures without functions, especially in DNA, has led to various functional suggestions, dealing in particular with the evolution or regulation of gene action, but evidence in favor of such a role still remains to be found. Other people have considered these sequences as parasitic DNA without any particular role in the economy of the host. It should be pointed out, however, that to say that no function is known does not imply that no function exists. The important question is: At what level should explanation be sought, and is it necessary? Furthermore, a piece of DNA that has spread without influence on the host phenotype may well have some secondary effect on the host. In particular, it may represent an evolutionary advantage for the host progeny. Such features as the fragmentation of structural genes into smaller DNA segments interspersed with intervening sequences and the presence, in many copies, of transposable elements able to spread through the genome and to carry DNA pieces from one place to another provide tools allowing transposition and, therefore, reassortment of coding fragments. Most of such rearrangements probably produce mere junk. Some, however, might result in the formation of a protein sequence able to perform, even inefficiently, some new cellular function. Further mutations would then allow the structure to be refined. Evolution has no foresight, and a genetic element cannot be selected because it might someday be of some help. Once it is there, however, whatever the reason, or absence of reason, for its presence, such a structure might prove "useful" and then become the target of some selective pressure on the host phenotype.

Biochemical novelty does not seem to have been a main driving force in the diversification of living organisms. The really creative part in biochemistry must have occurred very early. For the biochemical unity that underlies the living world makes sense only if most of the important molecular types found in organisms—that is, most of the metabolic pathways involved in the production of energy and in the biosynthesis and degradation of the essential building blocks—existed in very primitive organisms. Once this stage was passed, biochemical evolution continued as complexity developed. It was not, however, biochemical innovation that generated the diversification of organisms. In all likelihood, things worked the other way around. It was the selective pressure resulting from changes in behavior or in ecological niches that led to biochemical adjustments and changes in molecular types. What distinguishes a butterfly from a lion, a hen from a fly, or a worm from a whale is much less a difference in chemical constituents than in the organization and distribution of these constituents. The few really big steps in evolution clearly required the acquisition of new information. But specialization and diversification took place by using differently the same structural information. The studies on rates of evolution in frogs and mammals, for instance, suggest that sequence changes in structural genes occur, to a large extent, independently of anatomical changes. In contrast, regulatory changes, as manifested by studies of karyotypes and hybrid viability, appear to evolve in parallel with anatomical changes. Among neighboring groups, vertebrates for instance, chemistry is the same. As emphasized by Allan Wilson, differences between vertebrates are a matter of regulation rather than of structure.[9]

As already noticed by Karl Ernst von Baer at the beginning of the nineteenth century, among related organisms such as vertebrates the first steps of embryonic development are remarkably similar, divergences showing up only progressively as development proceeds. These divergences concern much less the actual structure of cellular and molecular types than their number and position. What makes a chicken wing and a human arm different is not so much the material out of which both are made, as the instructions specifying the way one or the other is built. Small changes modifying in time and space the distribution of the same structures are sufficient to affect deeply the form, function, and behavior of the final product: the adult animal. It is always a matter of using the same elements, of adjusting them, of altering here or there, of arranging various combinations to produce new objects of increasing complexity. It is always a matter of tinkering.

This is clearly illustrated by a comparison of human and chimpanzee macromolecules. Between these two species, gross differences in anatomy cannot be explained by differences in structural genes, which are amazingly similar. The average human protein chain is more than 99 percent identical with its counterpart in the chimpanzee. A large part of the differences in DNA sequences seems to result from redundancy in the genetic code or from variation in nontranscribed regions of DNA. For some fifty structural loci, the average genetic distance between chimpanzee and human turns out to be less than the average distance between sibling species barely distinguishable in morphology, and far less than the distance between any pair of congeneric species. As pointed out by Allan Wilson, the organismal differences between apes and humans must result mainly from genetic changes in a few regulatory systems.

A similar conclusion was reached by anatomists and paleontologists who have stressed the role of retardation of development as an evolutionary factor. Actually, some of the most dramatic events in evolution resulted from a change by which sexual maturity was achieved at an earlier developmental stage, so that previously embryonic characteristics were retarded in the adult, while previously adult characteristics were lost. It is now widely admitted that this process played an important role in the path that led to man. The development of the human embryo occurs within a matrix of retardation that retains a number of juvenile traits as compared with other primates and man's ancestors. This is clearly illustrated by the striking fact that humans resemble young apes much more than adult apes. Of course man is not descended from contemporary apes. Since the lineages of man and the apes have diverged, each has undergone further evolutionary changes by adapting to different types of life. Yet the common ancestor of man and apes resembled the latter more closely than the former. The retention of fetal patterns of gene expression during childhood probably made possible the evolution of such typically human features as small jaws and canine teeth, naked skin and upright posture. Furthermore, this retardation matrix with extended childhood appears to be closely associated with other marks of the hominization process, in particular with enlargement of the brain by prolongation of fetal growth, or socialization by strengthening family ties through the requirement of long parental care. The significance of delayed maturation for the hominization process has often been stressed. As recently pointed out by Stephen Gould, it is difficult to see how the distinctive set of human characteristics could have emerged outside the context of delayed development.[10] Diversification and specialization of organisms thus appear to result not so much from the appearance of new components as from a different use of the same components. Minor changes in regulatory circuits during development of the embryo can affect the rate of growth of different tissues or the time of synthesis of certain proteins, speeding up here, slowing down there.

Evolution is described in phylogenetic terms, that is, as differences between adult organisms. Yet differences between adult organisms merely reflect differences in the developmental processes that produce them. It is mainly through a net of developmental constraints that natural selection works by filtering actual phenotypes out of all possible genotypes. To really understand the way evolution proceeds, it is therefore necessary to understand embryonic development. Only with this knowledge will it become possible to evaluate what changes are compatible with the organization and the functioning of an organism, to define the rules and constraints of the evolutionary game. Unfortunately, to this day very little is known about embryonic development.

Biologists can describe with many details the composition of, say, a mouse. They can tell how the mouse moves around, how it breathes, how it digests. Yet they do not know how it forms itself out of an egg cell. A man is made of some 10^{13} and a mouse of some 10^{11} cells. Although all the cells that form an individual are direct descendants of the same single cell, the fertilized egg, they have different properties and perform different functions. It is often said that the chromosomes of the fertilized egg contain, coded in the linear sequence of the DNA, a description of the future adult. According to present views, however, what is coded in the chromosomes is the program to build that adult, that is, the instructions to manufacture its molecular structures and to put them into operation in time and space. However, the internal logic at work in the execution of the program remains completely unknown. It is generally admitted that a Laplacian demon able to examine the fertilized egg, its molecular structures and organization, would be able to describe the future adult. However, what kind of molecules besides DNA the demon would have to examine and what kind of algorithm it would have to use remain a complete mystery.

For the only logic that biologists really master is one-dimensional. As soon as a second dimension is added, not to mention a third one, biologists are no longer at ease. If molecular biology was able to develop so rapidly, this is largely because, in biology, information happens to be determined by simple linear sequences of building blocks, bases in nucleic acids and amino-acids in protein chains. Thus the genetic message, the relations between the primary structures, the logic of heredity, everything turned out to be one-dimensional. Everything was simple. No wonder molecular biologists do not like to abandon this type of work, preferring instead to continue in the analysis of a one-dimensional world with protein and DNA sequencing.

However, during the development of the embryo, the world is no longer merely linear. The one-dimensional sequence of bases in the genes determines in some way the production of two-dimensional cell layers that fold in a precise way to produce the three-dimensional tissues and organs that give the organism its shape, its properties, and, as Seymour Benzer puts it, its four-dimensional behavior.[11] How this occurs, however, is still a mystery. Biologists know in great detail the molecular anatomy of a human hand. They are completely ignorant of how the organism instructs itself to build that hand, what language it speaks to design a finger, what procedure it uses to sculpt a nail, how many genes are involved, and how these genes interact. Development and cell differentiation can be viewed as involving a series of binary decisions, the outcome of each decision determining the possibilities available for the following events. Sets of possibilities would thus be switched off at each fork. It is generally believed that such a process corresponds to a selective regulation of gene activity; but we do not even know the principles of the regulatory circuits that determine the number of cells, their distribution and movements, and the rates and direction of growth. We do not know the developmental tools of evolutionary tinkering.

Yet we have learned to imitate some of the natural processes and in particular to tinker with DNA in the laboratory. We have learned to cut and to knot DNA almost at will, to delete and to insert fragments. We know how to isolate certain structural genes, to mass-produce them and analyze their anatomy down to the last detail. All this work on recombinant DNA is in a way the triumph of our one-dimensional biology. It provides a new tool both for the study of fundamental problems and for many aspects of applied biology.

In order to produce a gene, say a human gene, in large quantities, it is necessary to insert it into the genetic makeup of a bacterium and then to grow this bacterium on a large scale. This type of work has aroused a lot of passion and hostility. It has been blamed for spoiling the quality of life and even for endangering human life. Actually it has become one of the major indictments of biology. Together with a series of contingencies such as research on fetuses, control of behavior, psychosurgery, or cloning of politicians, recombinant DNA is charged with providing biologists with the power to pervert both the body and the mind of humans. It is true that advances in science may be used for the better or the worse, that they may bring evils as well as benefits. Yet it is not science but interest and ideology that kill and enslave. In spite of Dr. Frankenstein and Dr. Strangelove, there are more evil priests and evil politicians than evil scientists. Harmful effects do not result only from events in which science is intentionally used for destruction. They may also come from distant and unpredictable consequences of actions intended to benefit humanity. Who would have predicted overpopulation as a result of medical advances? Or the scattering of virulent bacteria resistant to antibiotics as a result of the widespread use of these drugs? Or pollution as a result of the use of chemicals to improve crops? All are problems for which solutions have been or will be found.

With recombinant DNA, however, things simply happened the other way around. Apocalypse was predicted but nothing happened. If this work has raised endless debates, it is not so much because of the dangers that have been argued about—and which do not exceed those long recognized and mastered in the study of virulent bacteria or viruses—but because the idea that genes can be taken out of one organism to be inserted into the genetic makeup of another is by itself upsetting. The very notion of recombinant DNA is linked with the mysterious and the supernatural. It conjures up some of the old myths that have their origin in the deepest kind of human anxiety, the primitive terror associated with the hidden meaning of hybrid monsters, the revulsion caused by the idea of two beings unnaturally joined together.

For centuries, paintings of the Last Judgment have made much of frightful and impressive monsters, as illustrated for instance by the work of Hieronymus Bosch. The place of torment that Bosch represented as Hell is populated by the most horrible, the most terrifying monsters he could imagine. And these monsters are mainly unnatural hybrids. To endure what would seem to be among Hell's fiercest punishments, condemned sinners are left naked before such repugnant creatures as compounds of fish and rat, of dog and bird, or of insect and human; huge monsters crawling around their victims, swallowing them, running them through horrible mechanisms and torture devices; awful beasts eating, biting, ripping, clawing, flaying, and quartering. Such hybrid monsters presuppose a dislocation of the animal bodies followed by a reshuffling of the pieces, as if, in order to cause anxiety, Bosch was opposing the disorder of an antiworld to the harmony of our own world.

Work on recombinant DNA thus recalls many nightmares. It has the smell of forbidden knowledge. It calls up old myths in which mortals were harshly punished for having stolen a power exclusively reserved to the gods. Especially outrageous is the demonstration that it is so easy to fool about with the substance that is at the very basis of all life on this planet. Especially unforgivable the idea that one should consider as the result of some cosmic tinkering what remains the most baffling problem and the most amazing story: the formation of a human being; the process by which a sperm and an egg fuse, thus initiating the division of the egg cell, which becomes two cells, then four cells, then a small ball, then a small bag. And somewhere in this growing body, a small group of cells become individualized and multiply until they form a mass of billions of nerve cells. And it is with these cells that it becomes possible to learn, to speak, to read, to write, to count. It is with these cells that one is able to play the piano, to walk across a street without being run over by a car, or to come and lecture in Seattle. All these capacities are contained in our small mass of cells, all the grammar, all the syntax, all the geometry, all the music. And we do not have even the faintest idea of how it works. To me, this is the most amazing story that can be told on this planet, much more amazing than any novel or any science fiction.

TIME AND THE INVENTION OF THE FUTURE

Do not teach a monkey to climb trees.

—Shijing (*ancient Chinese* Book of Poetry)

One of the most attractive goddesses in Greek mythology is Eos, Aurora, the dawn. At the close of every night, rosy-fingered, saffron-robed Eos rises from her couch in the east, mounts her chariot drawn by the horses Lampus and Phaethon, and rides to Olympus where she announces the approach of her brother Apollo. Aphrodite was once vexed to find Ares, for whom she nursed "a perverse passion," in Eos' bed. She cursed Eos with a constant longing for young mortals. This is probably why Eos appears so attractive. From that time on, although she was married to Astraeus, Eos began secretly, and rather shamefully, to seduce young mortals, one after the other: first Orion, the son of Poseidon and one of the handsomest men alive; next Cephalos, who politely refused her advances, on the ground that he could not deceive his wife Procris. Eos metamorphosed him into another man who easily seduced Procris, so that Cephalos no longer felt any compunction about lying with Eos. Thereafter Eos carried off Cleitos, the grandson of Melanos, who was together the first mortal to be granted prophetic powers, the first to practice as a physician, and also, last but not least, the first to temper wine with water. Then Eos seduced both Ganymede and Tithonos, the two sons of King Tros who gave his name to Troy. Ganymede was considered the most beautiful youth alive. He was, therefore, chosen by the council of the gods to become the cupbearer of the great Zeus. While Ganymede was Eos' favorite lover, however, Zeus desired him as his own bedfellow; he disguised himself with eagle feathers and abducted Ganymede from Eos. By way of compensation, Eos begged Zeus to confer immortality upon her other lover, Tithonos. The request was granted by Zeus. Quite an unfortunate situation arose, however, once it was realized that, in her request for eternal life for her beloved, Eos had omitted to include also a request for eternal youth. Tithonos became older every day, grayer and more wizened. Worse, he talked incessantly with an ever shriller voice. Eventually rosy-fingered Eos got tired of nursing him. Once granted, however, immortality could not be canceled. Exasperated, Eos transformed Tithonos into a cicada and put him away in a box. If these two nightmares, death and aging, have to be dissociated, then the fate of Tithonos appears much worse than the reverse one, that of Dorian Gray, who is mortal but remains young while his portrait shows the marks of senescence.

The process of senescence is not understood. It is truly amazing that a complex organism, formed through an extraordinarily intricate process of morphogenesis, should be unable to perform the much simpler task of merely maintaining what already exists. Senescence is the postmaturational decline in the capacity to reproduce and survive that comes with aging. It consists, not in the alteration of a single system, but in a general deterioration of the whole body. A few decades ago, senescence was believed to result from a decay in the production of certain hormones, in particular of sexual hormones. Rejuvenation was then expected to follow implantation of young ape gonads in old people. Alas, the miracle was never achieved. Research on senescence has generally proceeded with the assumption that it will ultimately be explained by the alteration of one or a few physiological processes. This, however, appears more and more unlikely. Like other scientific fantasies such as the perpetual motion, the Fountain of Youth probably does not belong to the world of the possible.

The maximum duration of life is a characteristic of a species. It is, therefore, controlled by the genome. Senescence has even been considered as a stage of the developmental program, although this notion has never been defined. Once again, it was August Weismann who put senescence and what is often called "natural death" in an evolutionary perspective.[1] "I look at death as an adaptive phenomenon," he wrote, "because an infinite duration of the individual would represent a very inopportune luxury." It is, therefore, necessary that individuals should be continually replaced by new ones: "Worn out individuals are of no value for the species; they are even harmful since they take the place of those who are healthy." Weismann's argument has long been accepted. Yet, while he was right in discussing aging and death from an evolutionary perspective, he committed two sins. First, the argument is circular because to consider that old people are worn out and no longer able to reproduce is to start with precisely what has to be explained. Second, the selective mechanism is assumed by Weismann to act at the level of the species rather than that of the individual. But, as we have noted, natural selection cannot forecast the future in general or the fate of the species in particular. In Weismann's view, not only were the organisms subject to an inevitable decline, similar to the degradation of machines, but in addition a specific death mechanism had been designed by natural selection to eliminate old and, therefore, useless individuals. Senescence and mechanical wear, however, have very little in common, and despite decades of research, it has never been possible to show the existence of what could be called a death mechanism.

It is indeed difficult to figure out how a process that results in the shortening of life can be favored by natural selection. For if there is no specific death mechanism, one would imagine that slow deterioration of the organism should prevail over rapid deterioration. To avoid this paradox, Peter Medawar and George Williams have attempted to link senescence with the fact that selective pressure can operate only in the prereproductive period of the organism's life.[2] In every species, the most important individuals are those which reach sexual maturity, because they are the ones with the greatest capacity for propagation. Natural selection will, therefore, adjust the optimal state of animals to the time of their sexual maturity. In humans, for instance, maximal strength and resistance to disease is reached between twenty and thirty years, the lowest death rate being around fifteen. An animal thus appears to reach its optimum condition at the reproductive stage of its life cycle and to decline thereafter. Medawar and Williams have assumed the existence of genes that have harmful effects on the organism, either because of deleterious mutations or because of multiple effects, some beneficial, others detrimental. Natural selection would tend to accumulate these harmful effects in the postreproductive period of the animal's life, thus favoring deterioration of the body with age. In other words, vigor in youth should in a way be paid for by senescence. Selective forces that favor vigor in youth would increase the rate of senescence while other forces that postpone deleterious effects would decrease this rate. It would then be the balance between such opposite forces that would ultimately regulate the process of senescence and the maximum duration of life. For the present time, however, nothing is known of these hypothetical deleterious genes which remain abstract entities.

Aging gives life an inescapable temporal connotation. For the Greeks, time was punctuated by a variety of cyclical events and by the endless tide of life and death. "Men in their generations," said Homer, "are like the leaves of the trees. The wind blows and one year's leaves are scattered on the ground; but the trees burst into bud and put on fresh ones when the spring comes round. In the same way one generation flourishes and another nears its end."[3] This conception of an elusive destiny was applied to all of reality, determining the turnover of the seasons as well as the periodicity of celebration, and the succession of generations: cosmic time, religious time, and human time. Later in Greek history, time itself became deified as Chronos. In orphic cosmogony, for instance, Chronos stood at the very origin of the cosmos. It was represented as a kind of polymorphous monster and generated the primordial egg, which broke open to give birth to both the earth and the sky and, later on, to gods and mortals.[4]

In our own evolutionary mythology, time is also given a leading role. It is taken to be one of the factors that have shaped the world, and especially the living world. Indeed the requirement for time's arrow is one of the characteristic differences between biology and most aspects of physics. For, curiously enough, there is no arrow of time in the basic theories of physics. In the physical world, there are some asymmetries with respect to time, such as the expansion of the universe or the spherical electromagnetic waves that propagate outward from their source. Until rather recently, however, the fundamental laws of physics, quantum mechanics and electromagnetism, were considered time symmetric, and they are still believed to be nearly so. The birth and death of particles are processes that can be viewed as the strict reversal of each other. Asymmetry appears only in complementary phenomena. Until the recent description of an "irreversible thermodynamics," such a temporally asymmetric law as the second law was considered as only approximately true, this approximate truth being deducible from temporarily symmetric laws. Movies running backward make it possible to visualize what a world with reverse time might look like, a world in which milk would separate from coffee in the cup and rise upward into a milk jug; in which light rays would come out of the walls and converge into a sink instead of spreading from a source; in which a stone kicked out of the sea by innumerable and amazingly cooperative droplets of water would jump along a parabola to land in a human hand. In a world thus reversed in time, however, the processes in our brain and the formation of memory would also be reversed. So would past and future, and the world would appear just as it is.

In contrast to most aspects of physics, biology incorporates time as one of its essential parameters. The arrow of time can indeed be found throughout the whole living world, which results from an evolution in time. It can also be found in every single organism, which changes incessantly during its life. The future and the past represent totally different directions. Every creature moves from birth to death. Every individual's life is governed by development according to a plan, a feature which had a tremendous influence on Aristotle's philosophy and through it on all of Western culture, on its theology, its art, and its science. Molecular biology has now bridged the gap between this characteristic of living beings, development according to a plan, and the physical universe. The arrow of time required whenever life is involved has thus become part of our representation of the world. It is the specialty of biology, its stamp, so to speak.

Most organisms have internal clocks, which regulate their physiological cycles. All have memory systems on which their very existence, functioning, and behavior are based. The first one, the genetic system, is common to all organisms. It is the memory of the species, so to speak, and results from a billion years of evolution. It keeps, coded in DNA, a trace of the events that, generation after generation, have led to the present situation. As discussed in the previous chapters, the genes are not influenced directly by the life experience of the individual. Acquired characteristics are not transmitted to the progeny. Heredity does not learn from experience. If, after all, environment does influence heredity, it is always through the long twists and turns imposed by natural selection.

Complex organisms have evolved two other memory systems. The organization of both systems is under gene control and their function is to record certain aspects of the individual's life. The immune system was detected because the body keeps a memory of certain infections. It has long been known that some diseases do not occur twice in the same individual. As early as the fifteenth century, the Chinese inhaled a dried powder of smallpox crusts to gain immunity against smallpox. Three hundred years later, it was shown by Edward Jenner that inoculation of the related and benign cowpox could protect against subsequent infection with smallpox. However, the real start of immunology as a science came with Louis Pasteur when instead of inoculating a fresh culture of bacteria known to be able to kill a chicken in a few days, he accidently injected an aged culture of the same organism: not only did the chicken survive this injection but it turned out to have thus become immunized against further inoculation with the virulent culture.

A century later, the immune system has been found to attain an incredible degree of complexity. It involves several classes of very specialized cells, the lymphoid cells, which interact in various combinations, either by direct interaction between cells or by the intermediary of chemical signals. During development, the immune system learns to distinguish self from nonself. It thus becomes able to react either against self components that have become altered through pathological processes, or against the irruption of foreign molecules, the "antigens," in the body. The body reacts by producing, and secreting in the bloodstream, antibodies that neutralize the antigen, or through the action of specialized cells that directly bring about the destruction of the antigen, as in transplant rejection. In both cases, sets of cells with an enormous number of reactive potentialities are generated by a system in which fragments of genetic information in limited numbers are assembled in all possible combinations. In both cases, the potentially reactive cells are already available, only waiting to be activated by confrontation with an antigen. It is, therefore, the experience of the organism that actualizes immunological performances by selection among a large variety of preexisting structures.

The genetic system and the immune system thus function as memories to record the past of the species and of the individual respectively. A living organism, however, is not only the last link in an uninterrupted chain of organisms. Life is a continuous process which does not just recall the past, but also looks ahead. And the nervous system, which probably first evolved as a device to coordinate the behavior of various cells in multicellular organisms, then to record certain features of the individual's life, ultimately became able to invent the future.

—◇—

Time and the Invention of the Future

Living beings can survive, grow, and multiply only through a constant flow of matter, energy, and information. It is, therefore, an absolute necessity for an organism to perceive its environment or, at least, those aspects of its environment that are related to its life requirements. The simplest organism, the humblest bacterium, has to "know" what kind of food is available in order to adjust its metabolism accordingly. Such a microorganism's perceptions and reactions are under strict genetic control, each being limited to two alternatives. A bacterium can perceive in its environment only what its genetic program allows it to detect through a few proteins, each of which recognizes specifically a particular compound. For a bacterium, the external world is limited to a few substances in solution.

Obviously, the increase in performance that accompanies evolution requires a refinement of perception, an enrichment of the information received concerning the environment. There are many ways for living organisms to probe the external world. Some smell it; others listen to it; many see it. Every organism is so equipped as to obtain a certain perception of the outer world. Each species thus lives in its own unique sensory world, to which other species may be partially or totally blind. Honey bees, for instance, are blind to red light but are able to see ultraviolet light to which we are blind. A whole series of specific devices have evolved, such as ultrasonic echolocation in bats, electrosensitive organs in fishes, infrared eyes in snakes, sensitivity to polarized light in bees, and sensitivity to magnetic fields in birds. What an organism detects in its environment is always but a part of what is around. And this part differs according to the organism.

In lower vertebrates, sensory information is converted into motonervous information in a rigid way. Such animals appear to live in a world of global stimuli closely linked to processes of appropriate responses, what ethologists call "innate releasing mechanisms." In contrast, in birds and even more so in mammals, the enormous amount of information coming from the environment is sorted out by sense organs and processed by the brain, which produces a simplified and useful representation of the external world. The brain functions, not by recording an exact image of the world taken as a metaphysical truth, but by creating its own picture.

For each species, the world as it is perceived thus depends on the sense organs and on the way the brain integrates sensory and motor events. Even in those instances in which the same range of stimuli can be sensed, different species are wired to select particular features of the environment. Neural processing may separate the perceptual environment of different species as radically as if the stimuli they sense were coming from different worlds. We ourselves are so deeply entrapped in the representation of the world made possible by our own sense organs and brain, in other words by our genes, that we can barely conceive the possibility of viewing this same world in a different way. We can hardly imagine the world of a fly, an earthworm, or a gull.

No matter how an organism investigates its environment, the perception it gets must necessarily reflect so-called "reality" and, more specifically, those aspects of reality which are directly related to its own behavior. If the image that a bird gets of the insects it needs to feed its progeny does not reflect at least some aspects of reality, then there are no more progeny. If the representation that a monkey builds of the branch it wants to leap to has nothing to do with reality, then there is no more monkey. And if this did not apply to ourselves, we would not be here to discuss this point. Perceiving certain aspects of reality is a biological necessity; certain aspects only, for obviously our perception of the external world is massively filtered. Our sensory equipment allows us to see a tiger entering our room, but not the cloud of particles which, according to physicists, constitutes the reality of a tiger. The external world, the "reality" of which we all have intuitive knowledge, thus appears as a creation of the nervous system. It is, in a way, a possible world, a model allowing the organism to handle the bulk of incoming information and make it useful for its everyday life. One is thus led to define some kind of "biological reality" as the particular representation of the external world that the brain of a given species is able to build. The quality of such biological reality evolves with the nervous system in general and the brain in particular.

Some years ago, Harry J. Jerison of U.C.L.A. suggested that the quality of "biological reality" in relation to possibilities of behavior might actually have constituted an important factor of selective pressure for the development of the brain in mammals.[5] In this analysis he gave the concept of time a major role. During evolution the time parameter must have been progressively incorporated into the representation of the world since it could hardly operate in lower vertebrates. Among reptiles, for instance, there is little indication of time perception. Spatial representation is coded by an analyzer located in the retina itself. Early mammals were small animals confined to nocturnal life by the presence of large reptiles such as dinosaurs in the same areas. For the exploration of the environment at a distance, nocturnal life led to a replacement of vision by audition and smell. This had two consequences: first, an increase in the auditory area of the brain to harbor a new stock of neurons which could not be accommodated in the ear; second, a new way of processing spatial information according to a temporal code, somewhat like that of bats, which have a radar they use to detect objects by emitting a sound and locating its echo. Further steps would have later contributed to an enlargement of the brain and to an enrichment of the "biological reality" in mammals.

After the disappearance of the giant reptiles, when mammals could return to daylight life, they did not use the old reptilian visual system. Instead, they evolved a more refined system, with color vision and analyzers located no longer in the retina but in the brain. Visual and auditive information could thus be integrated through a coordinated spatial and temporal code with which it became possible to allocate light and sound stimuli to unique sources, that is, to individual objects that remain constant in time and space. If the brain of higher mammals can handle the tremendous amount of information coming in through the sense organs during wakefulness, it is because the information is organized in aggregates, in bodies that constitute the "objects" of the animal's spatio-temporal world, the very elements of its daily experience. Identification and perception of objects can thus be maintained despite changes in spatial and temporal perception.

It is possible to analyze in a similar way the encephalization steps that led to *Homo sapiens*, for they also involved an enrichment of the mental representation of the external world. Here again, according to Jerison, time should be given an important role, for the selective pressure that must have operated on hominids resulted in a need for the auditive perception of space, allowing for a more accurate location of sound-emitting sources. The information coming in through the different sense organs was integrated into a coherent picture of a spatio-temporal world, in which moving objects could be seen, heard, smelled, and touched, and in which, since the permanency of objects over time was ensured, their representation could be memorized. The organization of such a representation of the external world has certain implications with respect to time, in particular when two of the most fantastic powers of the brain are considered. On the one hand, memorized images of past events can be broken up into component parts and recombined to form hitherto unseen representations of new situations: this makes it possible not only to keep images of past events but also to imagine possible events and, therefore, to invent a future. On the other hand, the requirement for auditory perception of temporal sequences coupled with suitable changes in vocal sensory motor apparatus makes possible an entirely new way of symbolizing and coding such cognitive imagery. In this view, the role of language as a communication system between individuals would have come about only secondarily, as many linguists believe. Its primary function would rather have been, as with earlier evolutionary steps in mammals, the representation of a finer and richer "reality," a way of handling more efficiently a greater amount of information. As exemplified throughout the whole animal kingdom, communication can easily be established between individual organisms. Even among hominids which had to hunt and live in community, most of the information to be shared with others and concerning immediate features of life could be handled by means of rather simple codes. In contrast, to translate a visual and auditory world so that objects and events can be precisely labeled and recognized weeks or years later requires a much more elaborate coding system. The quality of language that makes it unique does not seem to be so much its role in communicating directives for action as its role in symbolizing, in evoking cognitive images. We mold our "reality" with our words and our sentences in the same way as we mold it with our vision and our hearing. And the versatility of human language also makes it a unique tool for the development of the imagination. It allows infinite combinations of symbols and, therefore, mental creation of possible worlds.

Acçording to such a viewpoint, each of us lives in a "real" world that is created by his brain with the information coming in through his senses and through language. This real world provides a stage where all the events of life take place. Although the experience to which the brain is exposed during a lifetime varies from one individual to another, the representation it creates is sufficiently similar to be communicable by one individual to another through words. Consciousness might be seen as the perception of self as an "object" placed at the center of "reality." The existence of the self as an object—that is, a person—is certainly one of the most compelling intuitions. It is difficult to decide at what stage of evolution it is possible to detect a beginning of self-awareness. One of the clues might be found in the capacity to recognize oneself in a mirror, a capacity which is considered to occur only at a certain level of complexity in the evolution of primates. Combined with the capacity for imagery, through which one recombines pieces of "reality" to re-create a representation of possible worlds in one's imagination, self-awareness provides a way of assessing the existence of a past that occurred before one's own life. It also enables one to imagine a time to come, to invent a future that contain's one's death and even a time after one's death. It allows a departure from the actual and the creation of a possible.

The old epistemological tradition, which is still in favor with many scholars, at least in Europe, was based mainly on introspection. It considered mental events to be of a different nature than physical events. Yet it seems very hard to imagine how an immaterial mind could have arisen from a process of evolution by natural selection. Endowing the elementary particles that constitute matter with some kind of a psyche does not help much, and the conclusion is inescapable that mind is a product of brain organization in the same way that life is a product of molecular organization. It is not clear whether we shall ever know how living organisms emerged from a lifeless universe, nor is it clear whether we shall ever understand the evolution of the brain and the emergence of this set of properties that we can hardly define but that we call mind.

Any evolutionary history of the brain and of the mind remains, therefore, merely a story, a scenario. A variety of scenarios can be, and have been, suggested which depend on the type of arguments—psychological, ethological, neurological, paleontological, and so forth—considered the most informative. The story told by Jerison is mainly derived from paleoneurological data, in particular from the relative brain and body sizes that can be studied in fossil vertebrates and enable one to determine the major steps in the encephalization process. This hypothesis appears particularly attractive because information about the outer world and the representation of reality constitutes a permanent factor of selection pressure throughout the evolution of the mammals, hominids included. An ever more refined representation of the external world, correlated with an increasing capacity to react to a variety of situations, should have resulted in an ever greater complexity of organisms in the "direction" generally attributed to evolution. Some human activities such as the arts, mythmaking, or the natural sciences may even be viewed as cultural developments in the same direction. The arts constitute, in a sense, efforts to communicate by various means certain aspects of a private representation of the world. Mythmaking aims, among other things, at integrating bits and pieces of information about the world into a coherent public picture. The natural sciences, which received a new impetus at the end of the Renaissance, represent an ancient way of refining the public representation of the world and providing a more comprehensive view of reality. All these activities call on human imagination. All operate by recombining pieces of reality to create new structures, new situations, new ideas. And a change in the representation of the world can result in a change in the actual physical world as evidenced by technological developments.

Most of what characterizes mankind is contained in the word "culture." The transmission of cultural features has a superficial analogy with that of biological features and is sometimes referred to as "cultural inheritance." The main similarity between the two systems is their tendency toward conservatism, mitigated by the possibility of change and, as a result, of evolution. Cultural features, however, spread by a Lamarckian mechanism. Cultural evolution can therefore occur at a rate which is orders of magnitude faster than that of biological evolution. Biologically, the human being of the twentieth century is apparently no different from the one who lived some 30,000 or 40,000 years ago. In contrast, the cultural, social, and technological world in which a human being at the end of this century is dying has little in common with the one in which he was born.

The more a scientific field deals with human affairs, the greater the chance that scientific theories will clash with traditions and beliefs; and also the more likely the data contributed by science will be distorted and used for ideological and political purposes. In recent years the old controversy about the relative contribution made by genes and by environment to the behavior of human beings has come up again. While in lower organisms, behavior is strictly determined by the genetic program, in complex metazoa the genetic program becomes less constraining, more "open" as Ernst Mayr puts it, in the sense that it does not lay down behavioral instructions in great detail but rather permits some choice and allows for a certain freedom of response.[6] Instead of imposing rigid prescriptions, it provides the organism with potentialities and capacities. This openness of the genetic program increases with evolution and culminates in mankind. A human being's forty-six chromosomes endow him with a whole series of physical and mental abilities which he can exploit and develop in a variety of ways depending on the environment and the society in which he grows and lives. For instance, it is his genetic makeup that gives a child his capacity to speak. But it is his environment that leads him to speak one particular language rather than another. Like every other character, the behavior of a human being is molded by a continual interplay of genes and environment.

Yet the close cooperation between biological and cultural determinants is often underestimated, if not simply denied, for ideological and political reasons. Instead of considering these two factors as complementary and indissolubly linked in the necessary cooperation that shapes human behavior, many people look upon them as conflicting forces and want to determine their relative importance—as if in the formation of human behavior and in its alteration, these two factors acted as opposing or mutually exclusive forces. In a series of debates, on education, on psychiatry, on sex differences, two extreme attitudes are confronted, for which the human brain appears, to use an analogy with musical media, either as a blank tape or as a phonograph record. A blank tape can be instructed by the environment to retain, and eventually play back, any piece of music, while a phonograph record, whatever the environment, is never able to play back anything but that particular piece which has been imprinted in its grooves.

The supporters of the blank tape theory are often influenced by Marxist ideology, according to which a particular individual is entirely molded by his class and education. For them the mental capacities of human beings have simply nothing to do with biology and heredity. Everything is necessarily a question of culture, of society, learning, conditioning, reinforcement, and mode of production. All biological diversity, all differences in the abilities and talents of individuals thus disappear. Nothing else matters but social and educational differences. The human brain is beyond biology! In this extreme form, however, this attitude is simply untenable. A scientific explanation of human behavior must essentially be of the same kind as that accounting for the behavior of, say, apes or dolphins. There is no learning without the bringing into play of a program that determines what can be learned, when, and under which conditions. The developing child passes through well-defined stages of learning. And the work of neurobiologists shows that the nervous circuits which underlie human capacities are, for a part at least, prewired at birth. In a sense, the supporters of the blank tape theory behave somewhat like the vitalists at the beginning of the nineteenth century. For the vitalists, living creatures could not obey the physical and chemical laws that govern the behavior of inanimate objects. They required something special, a mysterious vital force to account for their properties. Today, the vital force has disappeared. Just like inanimate objects, living organisms follow the rules of physics and chemistry. But they also obey additional laws; they must fulfill such requirements as nutrition and reproduction, which have no meaning in the inanimate world. Similarly, human beings are not subject only to biological laws. They also have to fulfill additional requirements, psychological, linguistic, cultural, social, economic, and so forth. It is not possible, therefore, to account for such a complex system as the human brain in terms of a single discipline or even a series of sketchy accounts from different disciplines, the relative importance of each one being acknowledged by a particular weighting coefficient. Just as biology relates to physics, the study of man can neither be reduced to biology nor do without it.

No less untenable appears, therefore, the opposite attitude, that of the genetic phonograph record, a viewpoint which is generally associated with conservative philosophy and underlies various forms of racism and fascism. This conception ascribes to the genetic makeup of a human being almost the whole of his mental abilities and deserts. It denies any importance to his environment or upbringing, thus ruining any hope of improvement by training and learning. As long as the world was considered a product of creation, "human nature" was viewed as a part of the general harmony of the universe. It was a gift of God, who had endowed mankind with a set of properties and had fixed the rules governing the conduct of human affairs according to a well-defined social, economic, and political hierarchy. Once evolution was substituted for creation, the defenders of the social status quo had to replace God's will by something. Biological constraints were then invoked as a scientific guarantee, setting the limits of human behavior. For if the actual performances of an individual merely express his genetic potentialities, then social inequalities result from biological inequalities and it is pointless to even think of modifying the social hierarchy. In its modern version, this conception of the genetic recording looks for support in two areas. The first one is the kind of reductionism that is favored by some of the more simpleminded sociobiologists who want to explain the human mind as a machine genetically programmed down to the last detail. The second area to which the extreme hereditarian viewpoint looks for support is the study of IQ determinism, as deduced, for instance, from comparing the scores of monozygotic and dizygotic twins. The meaning of IQ tests, what they measure, the possibility of designing culture-free tests, all this is still the theme of passionate controversies. Without entering the debate, I would just like to point out the amazement caused in the naive biologist by the very principle of IQ tests. How can one hope to quantify what is usually called general intelligence—that is, something we cannot succeed in defining, and which includes such a variety of elements as the representation of the world and its underlying forces, the capacity to react to a variety of circumstances under a variety of conditions, broadness of outlook, quickness in apprehending all the components of a situation and making a decision accordingly, the capacity to evaluate the consequences of a decision, the ability to detect similarities that are more or less hidden, or to compare what *a priori* does not seem comparable, and so on? How can one hope to quantify such a heterogeneous set of complex properties using a single parameter moving linearly on a scale from 50 to 150? As if the important thing in science were to measure, whatever it might be that is measured. As if, in the dialogue between theory and experiment, the latter came first. Such a belief is simply wrong. In science, it is always the theory that initiates the process. Experimental data can be obtained and their meaning seen only in the frame of the theory. The emotional character of the nature-nurture controversy is further illustrated by recent findings concerning what supporters of the extreme hereditarian viewpoint have long considered to be one of their strongest arguments, namely, the findings of the British psychologist Cyril Burt on the IQ of twins. These data turned out, at least partly, to have been faked.[7]

Present-day biology actually has little to say about human behavior and the genetic component of mental abilities. Genetics proceeds from what is visible—the observable characteristics, the phenotype as it is called—to deduce what is hidden—the state of the genes, what is called the genotype. All the knowledge about heredity harvested by classical genetics or molecular biology is based on this method. This procedure works well when the phenotype reflects more or less directly the genotype, as in the case, for instance, of blood groups; or of some malformations which may be followed through successive generations; or even of certain diseases which appear in some way related to the genetic makeup of individuals. Most frequently the latter relation is not so much a complete correlation, but rather a probability: under similar conditions of life, a particular type of cancer or arthritis is more likely to occur among those people who possess certain genes than among others. The procedure of genetics, however, does not work in the study of mental abilities. In principle, an artificial selection experiment might be set up and heritability estimated. Artificial selection, however, is not feasible in humans. Furthermore, intellectual performance as observed in an individual does not directly reflect the state of his genes. It reflects the state of the many structures that intervene between the genotype and the phenotype; structures hidden in the depth of the brain, which function at many levels of integration, but to which there is presently no experimental access. No doubt the development of these brain structures is under some genetic control, as shown by the deficit in human performance resulting from certain mutations or chromosomal anomalies. No doubt, also, it is just as much under some environmental control, as shown by the deficit in a child's affective and cognitive performance resulting from lack of attention and love.

Any normal child at birth possesses the ability to grow up in any community, to speak any language, to accept any religion, any social convention. Most likely, the genetic program lays down what might be called *accommodation structures* which allow the child to react to external stimuli, to expect and notice regularities, to memorize them and reassort their elements into new combinations. By learning, these nervous structures become progressively more refined and more elaborated. It is through the constant interplay of biological and environmental factors during the development of the child that the nervous structures underlying mental abilities can mature and become organized. To attribute a fraction of the final organization to heredity and the remaining part to environment is simply pointless—as pointless as asking whether the inclination of Romeo toward Juliet has a genetic or a cultural basis. Like every living organism, a human being is genetically programmed, but he is programmed to learn. Many possibilities are offered at birth by nature. What becomes actualized is progressively developed during life by interaction with the environment.

The diversity generated by sexual reproduction among individuals in a human population is seldom seen as what it is: one of the main forces driving evolution, a natural phenomenon without which we should not be here. Most frequently this diversity is viewed either as a subject of scandal by those who criticize the social order and want to make all individuals equivalent or as a means of domination by those who want to justify this social order by an alleged natural order in which all individuals are graded as a function of the "norm," i.e., themselves. In spite of many claims, it is not science that determines politics, but rather political attitudes that distort and misuse science to find self-confirmation. In many cases, a confusion is more or less consciously entertained between two quite distinct notions: identity and equality. The former refers to the physical and mental properties of individuals, the latter to their social and legal rights. One is a matter for biology and education, the other for ethics and politics. Equality is not a biological concept. Two molecules or two cells are never said to be equal. Nor are animals, as George Orwell pointed out. Obviously, it is the social and political aspect that is at stake in the controversy, with the desire either to base equality on identity or to justify inequality by diversity. As if equality had not been invented precisely because human beings are not identical. If they all were as similar as monozygotic twins, the very concept of equality would be worthless. Its value and its importance come just from the diversity of individuals, from their differences in the most varied domains. Diversity is one of the great rules in the biological game. All along generations, the genes that constitute the inheritance of the species unite and dissociate to produce those ever fleeting and ever different combinations: the individuals. And this endless combinatorial system which generates diversity and makes each of us unique cannot be overestimated. It gives the species all its wealth, all its versatility, all its possibilities.

Diversity is a way of coping with the possible. It acts as a kind of insurance for the future. And one of the deepest, one of the most general functions of living organisms is to look ahead, to produce future as Paul Valéry put it.[8] There is not a single movement, a single posture that does not imply a later on, a passage to the next moment. To breathe, to eat, to move is indeed to anticipate. To see is to foresee. With each of our actions and each of our thoughts we are engaged in what will be. An organism is living insofar as it is going to live, even if only for a short while.

Selection from preexisting diversity appears as the means most frequently used in the living world to face an unknown future: the short-term future, with diversity at the molecular level, as observed with induced enzyme synthesis in bacteria or antibody production in vertebrates; the long-term future, with the diversity of species—whose incredible number allows life to settle in the most diverse areas and under the most extreme conditions on this planet—and above all with the diversity of individuals, which represent the main target of natural selection. If we were all equally sensitive to a virus, the whole of mankind could be wiped out by a single epidemic. We are 4.5 thousand million unique individuals so as to face possible hazards. It is the uniqueness of the person that makes the idea of producing perfect replicas by cloning so revolting.

In humans, natural diversity is further strengthened by cultural diversity, which allows mankind to better adapt to a variety of life conditions and to better use the resources of the world. In this area, however, we are now threatened with monotony and dullness. The extraordinary variety which humans have put into their beliefs, their customs, and their institutions is dwindling every day. Whether peoples die out physically or become transformed under the influence of the model provided by industrial civilization, many cultures are disappearing. If we do not want to live in a world covered with a single technological, pidgin-speaking, uniform way of life—that is, in a very boring world—we have to be careful. We have to use our imagination better.

Our imagination displays before us the ever changing picture of the possible. It is with this picture that we incessantly confront what we fear and what we hope. It is to this possible that we adjust our wishes and our loathings. Yet, while it is part of our nature to produce a future, the system is geared in such a way that our predictions have to remain dubious. We cannot think of ourselves without a following instant, but we cannot know what this instant will be like. What we can guess today will not be realized. Change is bound to occur anyway, but the future will be different from what we believe. This is especially true in science. The search for knowledge is an endless process and one can never tell how it is going to turn out. Unpredictability is in the nature of the scientific enterprise. If what is to be found is really new, then it is by definition unknown in advance. There is no way of telling where a particular line of research will lead. This is why it is not possible to select some parts of science and to reject others. As pointed out by Lewis Thomas, either you have science or you don't have it.[9] And if you have it you cannot take only what you like. You have to accept as well the unexpected and disturbing results.

In this book, I have tried to show that the scientific attitude has a well-defined role in the dialogue between the possible and the actual. The seventeenth century had the wisdom to introduce reason as a useful and even necessary tool for handling human affairs. The Enlightenment and the nineteenth century had the folly to consider it to be not merely necessary but even sufficient for the solution of all problems. Today, it would be still more foolish to decide, as some would like, that because reason is not sufficient, it is not necessary either. Yet, while science attempts to describe nature and to distinguish between dream and reality, it should not be forgotten that human beings probably call as much for dream as for reality. It is hope that gives life a meaning. And hope is based on the prospect of being able one day to turn the actual world into a possible one that looks better. When the French writer Tristan Bernard was arrested with his wife by the Gestapo, he told her: "The time of fear is over. Now comes the time of hope."

NOTES

MYTH AND SCIENCE

1. G. L. de Buffon, *Oeuvres complètes*, vol. 3, *Histoire des animaux* (Paris: Imprimeries Royales, 1774).

2. A. Weismann, "La reproduction sexuelle et sa signification pour la théorie de la sélection naturelle," in *Essais sur l'hérédité* (Paris: C. Reinwald et Cie, 1892).

3. R. A. Fisher, *The Genetical Theory of Natural Selection* (Oxford: Oxford University Press, 1930); H. J. Muller, "Some Genetic Aspects of Sex," *Amer. Naturalist* 66 (1932): 118-38; G. C. Williams, *Sex and Evolution* (Princeton: Princeton University Press, 1975); J. Maynard Smith, *The Evolution of Sex* (Cambridge: Cambridge University Press, 1978).

4. P. B. Medawar, *The Hope of Progress* (New York: Doubleday, 1973).

5. W. Paley, *Natural Theology*, vol. 1 (London: Charles Knight, 1836).

6. J. Lederberg, *J. Cell. Comp. Physiol.* 52 (1958, suppl. 1): 398.

7. A. Weismann, "La prétendue transmission héréditaire des mutilations," in *Essais sur l'hérédité* (Paris: C. Reinwald et Cie, 1892).

8. G. C. Williams, *Adaptation and Natural Selection* (Princeton: Princeton University Press, 1966).

9. S. J. Gould and R. C. Lewontin, "The Spandrels of San Marc and the Panglossian Paradigm: A Critique of the Adaptationist Programme," *Proc. R. Soc. London* B 205 (1979): 581-98.

10. Author's translation of passage from Voltaire, *Candide*, in *Romans et contes* (Paris: Gallimard, La Pléiade, 1954).

11. N. Chomsky, *Problems of Knowledge and Freedom: The Russell Lectures* (New York: Pantheon Books, 1971).

EVOLUTIONARY TINKERING

1. J. Fernel, "De abditis rerum causis," in *Opera*, vol. 1 (Geneva, 1637).

2. A. Paré, *Oeuvres complètes*, vol. 1, *Le premier livre de l'anatomie* (Paris, 1840).

3. See R. E. Dickerson, "Cytochrome c and the Evolution of Energy Metabolism," *Scientific American* 242 (1980): 136-53.

4. G. G. Simpson, "How Many Species?" *Evolution* 6 (1952): 342.

5. C. Levi-Strauss, *La pensée sauvage* (Paris: Plon, 1962).
6. C. Darwin, *The Various Contrivances by which Orchids are Fertilized by Insects* (New York: D. Appleton, 1886); M. Ghiselin, *The Triumph of the Darwinian Method* (Berkeley: University of California Press, 1969).
7. E. Mayr, "From Molecules to Organic Diversity," *Fed. Proc. Am. Soc. Exp. Biol.* 23 (1964): 1231-35.
8. P. McLean, "Psychosomatic Disease and the Visceral Brain," *Psychosom. Med.* 11 (1949): 338-53.
9. M. C. King and A. C. Wilson, "Evolution at Two Levels in Humans and Chimpanzees," *Science* 188 (1975): 107-16.
10. S. J. Gould, *Ontogeny and Philogeny* (Cambridge, Mass.: Harvard University Press, 1977).
11. S. Benzer, "The Genetic Dissection of Behavior," *Scientific American*, December 1973, pp. 24-37.

TIME AND THE INVENTION OF THE FUTURE

1. A. Weismann, "La durée de la vie," in *Essais sur l'hérédité* (Paris: C. Reinwald et Cie, 1892).
2. P. B. Medawar, *The Uniqueness of the Individual* (New York: Basic Books, 1957); G. C. Williams, "Pleiotropy, Natural Selection and the Evolution of Senescence," *Evolution* 11 (1957): 398-411.
3. Homer, *The Iliad*, trans. E. V. Rieu (London: Penguin Classics, 1966).
4. See J. P. Vernant, *Mythe et pensée chez les Grecs* (Paris: Maspero, 1971).
5. H. J. Jerison, *Evolution of the Brain and Intelligence* (New York: Academic Press, 1973).
6. E. Mayr, "The Evolution of Living Systems," *Proc. Nat. Acad. Sci. U.S.* 51 (1964): 934-41.
7. L. J. Kamin, *The Science and Politics of IQ* (Hillsdale, N.J.: Erlbaum, 1974).
8. P. Valéry, Oeuvres I (Paris: Gallimard, La Pléiade, 1962).
9. L. Thomas, *The Medusa and the Snail* (New York: The Viking Press, 1979).

ABOUT THE AUTHOR

François Jacob was born in 1920 in Nancy, France. After attending the Lycée Carnot in Paris, he began studying medicine at the Faculty of Paris with the intention of becoming a surgeon. In June 1940 his studies were interrupted by the war; he left France to join the French Free Forces in London and fought with the Allies in northern Africa, where he was severely wounded. For his war service he was awarded the Cross of the Liberation, one of France's highest decorations.

After the war, François Jacob finished his medical studies, completing his doctorate in 1947. Unable to practice surgery because of his war injuries, he worked in various fields before turning to biology. In 1950 he entered André Lwoff's laboratory at the Institut Pasteur. He obtained his doctorate in science in 1954 at the Sorbonne. In 1956 he was appointed Laboratory Director of the Institut Pasteur, then in 1960 head of the Department of Cell Genetics. In 1964 he was also appointed Professor of Cell Genetics at the Collège de France. He is a member of the French Academy of Sciences, the Royal Danish Academy of Sciences and Letters, the American Academy of Arts and Sciences, the National Academy of Sciences of the United States, the American Philosophical Society, the Royal Society (London), and the Royal Academy of Medicine (Belgium).

François Jacob has worked mainly on the genetic mechanisms of bacteria and bacterial viruses, on the information transfer and regulatory system in the bacterial cell. In 1965 he was awarded jointly with André Lwoff and Jacques Monod the Nobel Prize for Medicine.

In recent years, François Jacob has been studying the mouse embryo and mouse teratocarcinoma, a tumor originating from germ line cells, with the aim of analyzing early stages of development in mammals. He has also been long interested in the origins of ideas in biology, and has written a book on the history of heredity, *The Logic of Life* (reissued in 1982 by Pantheon, with a new preface by the author). His excursions into the philosophy of science form the basis of these Danz lectures.